UNDERSTANDING ELECTRONICS
With 32 Working Circuits

UNDERSTANDING ELECTRONICS
With 32 Working Circuits

by

R. H. WARRING

TAB BOOKS Inc.

BLUE RIDGE SUMMIT, PA. 17214

FIRST PRINTING—JANUARY 1979
SECOND PRINTING—FEBRUARY 1980
THIRD PRINTING—OCTOBER 1980
FOURTH PRINTING—MARCH 1981

Printed in the United States of America

Understanding Electronics was originally published in 1978.
Printed by permission of Lutterworth Press.

Library of Congress Cataloging in Publication Data

Warring, Ronald Horace.
    Understanding electronics.

    Includes index.
    1. Electronics—Amateurs' manuals. I. Title.
TK9965.W39   621.38   78-26492
ISBN 0-8306-9845-0
ISBN 0-8306-1113-4 pbk.

Contents

Introduction

If you find electronics difficult to understand — then you are one of a vast majority! One good reason for this is that you cannot 'see' how electronic circuits work — only switch them on and hope that they do work. So home construction of electronic circuits may seem something of a gamble, not helped by the fact that the 'plans' are usually in the form of theoretical circuit diagrams with components designated by symbols instead of their physical shape and size, and quite probably in nothing like the arrangement they will appear on a practical working circuit.

Take heart at this point! In at least nine cases out of ten, if a circuit built to a published 'plan' does not work it will be for one of two *simple* reasons: either poorly made connections or wrong connections. The only 'mystery' about that is that it happens — even to experts.

Electronic circuits are about components and how they behave when connected together in various ways. This book sets out to explain just that, in simple easy-to-understand fashion, without any difficult mathematical calculations or 'theoretical' explanations. What they are, what they look like, what they do, and what sort of circuits they are used in. A basic groundwork in the practical side of simple electronics — and the subject can be quite simple. And to make the subject more 'realistic', there are thirty-two suggested *working* circuits to build, or experiment with.

There is still the question of how to 'read' and understand circuit drawings, and turn them into practical working circuits, so separate chapters are devoted to these two particular subjects. There is also a further chapter on printed circuit construction.

Since most simple electronic circuits are battery powered, the subject of different battery types and their performances is also covered in some detail. Also circuits for building a battery charger, either to operate off the mains or from a car battery.

Understanding Electronics should be an excellent 'starter' for anyone wanting to take up practical electronics as a hobby interest — and an ideal reference to back up, and make it easier to understand, the types and construction of more elaborate circuits described in rather more advanced books, monthly journals, and so on. It should prove invaluable as a source of information which has been left out of other electronic books on the author's assumption that 'everyone should know that'; the reason, in fact, why such books are so often described as 'too technical' or 'too advanced'. So this present book should fill that particular gap.

LIST OF WORKING CIRCUITS TO BUILD AND/OR EXPERIMENT WITH

CHAPTER 1

Units, Abbreviations and Symbols

In electronics there are seven basic *units* to measure quantities which define what is going on in a circuit. These are (together with the letter symbols used as abbreviations):

Volts (V) — a measure of the potential difference, *emf* (electromotive force), or *voltage* in a circuit. For practical purposes, potential difference, *emf* and voltage all really mean the same thing.

Amps (A) — a measure of the *current* flowing in a circuit.

Watts (W) — a measure of the *power* developed by the flow of current through a circuit.

The other four refer to the effect of components in the circuit, viz:

Ohms (Ω) — a measure of the resistance or individual resistances in a circuit when the current flow is direct *(dc)*.

Impedance (Z) — a measure of the effective resistance or individual resistances in a circuit when the current flow is alternating *(ac)*.

Farads (F) — a measure of the *capacitance* present in a circuit or produced by individual components, i.e. capacitors.

Henrys (H) — a measure of the *inductance* present in a circuit or produced by individual components such as coils.

Reactance (X) — the combined effect of inductance and capacitance in an *ac* circuit.

Capital letters are also used as abbreviations for voltage and current. Strictly speaking E (for *emf*) is the correct symbol for a voltage source, with V (for volts) in other parts of the circuit. V_s can be used instead of E for a source voltage. The capital letter I is used for current. In some circuits lower case letters are used to indicate voltages and currents flowing in different parts of a circuit, e.g. v and i, respectively. These may have a reference annotation attached, particularly in the case of transistor circuits, e.g. v_e describing emitter voltage.

The relationship between units is explained in Chapter 3. There are also various other units employed in electronics, the use and meaning of which will be made clear in appropriate chapters.

In practical circuits, numerical values of these units may be very large, or very small. Resistance values, for example, may run to millions of ohms. Capacitor values may be in millionths or even million-millionths of a farad. To avoid writing out such values in full, prefixes are used to designate the number of noughts associated with the particular value involved. Again the symbol rather than the full prefix is normally used:

mega (M) — meaning *times* 1,000,000
kilo (k) — meaning *times* 1,000
milli (m) — meaning divided by 1,000 (or 1/1,000th)
micro (μ) — meaning divided by 1,000,000 (or 1/1,000,000th)
nano (n) — meaning divided by 1,000,000,000 (or
                                    1/1,000,000,000th)
pica (p) — meaning divided by 1,000,000,000,000 (or
                                    1/1,000,000,000,000)

For example, instead of writing out 22,000,000 ohms in full, this would be shown as 22 Mohms, or more usually 22 MΩ, using symbols both for the prefix and basic unit. Similarly a capacitor value of 0.000,000,000,220 farads would be shown as 220 picaF, or more usually 220 pF.

The *multipliers* (M and k) are most commonly associated with values of *resistors*, and also for specifying radio frequencies. The lowest *divisor* (m) is most usually associated with the values of *current* typical of transistor circuits, etc. It is also used to specify most practical values of inductances. The larger *divisors* (μ, n and p) are most commonly associated with capacitor values.

More Abbreviations
Single capital letter abbreviations are also used for components. The main ones are:

C — for capacitors
D — for diodes
L — for coils
R — for resistors

These are all standard and universally accepted abbreviations. With other components this is not always the case. Thus *transistors* may be designed T, TR, Tr, VT or even Q on circuits originating from different sources. The use of TR (or Tr) is preferred, leaving the letter T as the abbreviation for transformers. But note the abbreviation FET (or fet) is always used for a *field effect transistor*.

In practical circuits more than one of the same type of components are normally used. Individual components of the same type are then designated by numbers (usually reading from left to right across the circuit) associated with the component *symbol* (Fig. 1) (*see also* Chapter 3). Thus resistors would be designated R1, R2, R3, etc; capacitors C1, C2, C3, etc . . . and so on. There is no 'correct' or specific sequence in which such numbers are allocated. They are there only to identify a particular component.

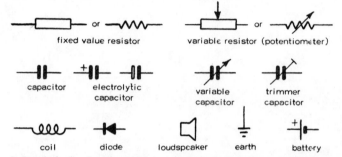

Fig. 1 Symbols for basic circuit components. Other symbols are given in later chapters

Here are some other general abbreviations which are widely used, although again they may be shown in various different ways — capital letters, or lower case letters in upright or italic, with or without full stops. Thus the abbreviation of alternating current may appear in five different ways:

AC   a.c.   *a.c.*   ac   *ac*

The general preference is that all such abbreviations should be in lower case italic without full stops, and so the following abbreviations are shown that way:

ac — alternating current
af - audio frequency
agc — automatic gain control
am* — amplitude modulated (or amplitude modulation)
dc — direct current
eht — extra high tension
fm* — frequency modulated (or frequency modulation)

* There is a good case for retaining *capital* letters for these abbreviations as AM and FM radios are widely quoted in this way.

hf — high frequency
ht — high tension
ic — integrated circuit
if — intermediate frequency
lf — low frequency
rf — radio frequency
uhf — ultra high frequency
vhf — very high frequency.

CHAPTER 2

dc and *ac*

A basic direct current (*dc*) circuit is simple enough to understand. A source of electrical force (e.g. a battery) is connected via wires to various components with a path to the source. Current then circulates through the circuit in a particular direction. Fig. 2 shows a very elementary circuit of this type where a battery is connected to a *dc* electric motor and is compared with a similar closed loop hydraulic motor in a simple recirculating system.

It is obvious what happens in the *hydraulic* circuit. The pump is a source of pressurized water which impinges on the vanes of the hydraulic motor to drive it. There is a flow of water around the system. At the same time there will be some loss of pressure energy due to friction of the water flowing through the pipes and the motor. This is the resistance in the circuit. But most of the pressure energy delivered by the pump is converted into *power* by the hydraulic motor.

In the electrical circuit counterpart, the battery is a source of *electrical* pressure (which in simple terms we designate *voltage*). This forces an electrical *current* to flow through the circuit, opposed by the *resistance* offered by the wiring and the electric motor coils. Again most of the original electrical energy in the battery is converted into *power* by the electric motor. Provided the battery voltage does not change, a constant value of current will flow through the circuit always in the same direction and the electric motor will continue to run at a constant speed.

Conventionally, *dc* current flow is regarded as being from 'positive'

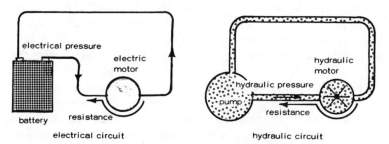

Fig. 2. An electrical circuit is similar to a hydraulic circuit

to 'negative' of a battery or any other *dc* source (such as a dynamo). It is a 'stream' flow, just like the water flow in the hydraulic circuit, but the 'stream' is actually composed of sub-atomic particles or *electrons*. Unfortunately, after convention had established the 'positive to negative' flow definition it was found that this electron stream flow was actually from *negative to positive*. This does not matter for most practical purposes, but for an understanding of how transistors and other solid-state devices work it is necessary to appreciate this 'reverse' working.

'Positive' always seems stronger than 'negative', so it is difficult to think of current as flowing other than from positive to negative. We can relate this to electrons flowing from negative to positive by thinking of electrons as particles of negative electricity. Being 'weaker' (negative), they represent a 'reverse flow', setting up conditions for a *positive* flow of current — from *positive to negative*. Otherwise, simply forget the difference and work on the practical fact that + and − are only terms of convenience used to ensure that components in a circuit which have positive and negative sides are connected up the right way round! This applies mainly to batteries, transistors, diodes and electrolytic capacitors.

All materials are composed of atoms in which there is a stable balance of positive and negative charges (except in the atoms of radio-active elements). The application of an electrical pressure will cause electrons to be displaced from the atom, leaving it with an effective *positive* charge. It is then in a state to attract any stray electrons. Since there is electrical *pressure* present, this means that there will be a movement of electrons along the chain of atoms comprising the wiring and component(s) in the complete return circuit. It is this movement that constitutes the electric current flowing through the circuit, the strength of the current being dependent on the number of electrons passing any particular point in the circuit in a given time. Break the circuit and the 'pressure' is broken, so current flow ceases. So, in fact, the analogy with a hydraulic circuit is not really valid in this instance (the hydraulic pump will still deliver water under pressure if its circuit is broken until it has emptied the fluid in the circuit between the break and the pump).

Atoms of materials like metals will give up electrons readily when subject to electrical pressure, and so make good *conductors* of electricity. Atoms of most non-metals, including plastics, are reluctant to give up electrons even under high electrical pressure, and so are essentially non-conductors. If extremely resistant to giving up atoms, they are classified as *insulators*.

Summarizing, then, a *dc* circuit when connected or switched on provides a constant flow of current in one direction through the circuit Fig. 3 — unless something changes in the circuit (e.g. source voltage changes, or a circuit resistance value alters). The value of this current is determined by the source *voltage* and the total *resistance* in the circuit (*see* Chapter 3). Current flow is also regarded as *positive* (or positive current).

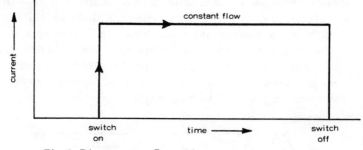

Fig. 3. Direct current flow with constant circuit resistance

In the case of an *ac* circuit the source of electrical pressure continually reverses in a periodic manner. This means that current flows through the circuit first in one direction (positive) and then the other (negative). In other words, a simple graph of current flow with time will look like Fig. 4. The 'swing' from maximum positive to maximum negative is known as the *amplitude* of the *ac* current. Also one complete period from zero to maximum positive, back to zero, down to maximum negative and back to zero again is known as a *cycle*.

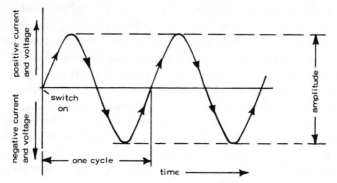

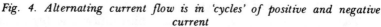

Fig. 4. Alternating current flow is in 'cycles' of positive and negative current

These cycles may occur at varying rates from a few times a second to millions of times a second and define the *frequency* of the *ac* current, frequency being equal to the number of cycles per second. In the case of the domestic mains supply (in Britain), for example, the frequency is 50 cycles per second. But 'cycles per second' is an obsolete term. It is now called Hertz (abbreviated Hz). Thus standard mains frequency is 50 Hz.

Apart from the fact that *ac* is continually swinging from positive to negative current flow, the other difference compared with *dc* is that the actual current value present is also changing all the time. It does, however, have an 'average' value which can be defined in various ways. The usual one is the Root Mean Square (or *rms*) which is equal to 0.7071 times the maximum cycle values for sine wave *ac* such as normally generated by an alternator — Fig. 5. Alternating current may, however, be generated with other types of waveform.

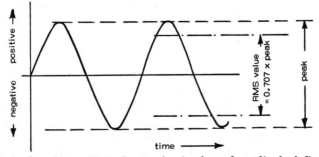

Fig. 5. Peak and Root Mean Square (rms) values of amplitude defined and compared

Another characteristic of *ac* is that both the voltage *and* current are continually changing in similar cycles. Only rarely, however, will the voltage and current both attain maximum and zero values at exactly the same time. In other words the current (waveform) curve is displaced relative to the voltage (waveform) curve — Fig. 6. This displacement is known as a *phase difference*. It is normally expressed in terms of the ratio of the actual displacement to a full cycle length on the zero line, multiplied by 360 (since a full cycle represents 360 degrees of *ac* working). This is called the *phase angle*. Usually the current will 'peak' after the voltage (i.e. be displaced to the right on the diagram), whereupon the current is said to be lagging and the phase angle is referred to as *angle of lag*.

The use of the term 'angle' can be a bit confusing at first. It is

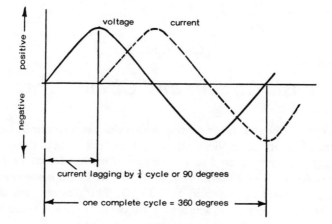

Fig. 6. *Current usually 'lags' behind voltage in an alternating current circuit*

really a matter of mathematical convenience, useful in more complicated calculations involving vector diagrams. For a general understanding of *ac* it is better to think of 'angle' as meaning a particular 'number point' on a line length representing one full cycle divided into 360 divisions. Thus a phase angle of 30 degrees can be understood as a point 30/360ths *along* that line.

Phase difference (phase angle) can be an important factor in the design and working of many alternating current circuits because when a current 'lags' (or 'leads') the voltage, the 'timing' aspects of a circuit are affected.

CHAPTER 3

Basic Circuits and Circuit Laws

As noted in Chapter 2, the current which flows in a simple *dc* circuit is dependent on the applied voltage and the resistance in the circuit. Voltage can be measured directly by a *voltmeter* placed across the battery (or *dc* source) terminals; and *current* by an *ammeter* connected in *series* in the circuit, as in Fig. 7. This diagram also shows the circuit components in symbolic form (in the case of a resistor this can be drawn as a plain rectangle or a zig-zag line).

The relationship between voltage (E), current (I) and resistance (R) is given by Ohm's Law:

$$I = \frac{E}{R}$$

In plain language:

$$\text{amps} = \frac{\text{volts}}{\text{resistance in ohms}}$$

or the formula can be rewritten:

$$\text{volts} = \text{amps} \times \text{ohms}$$

$$\text{ohms} = \frac{\text{volts}}{\text{amps}}$$

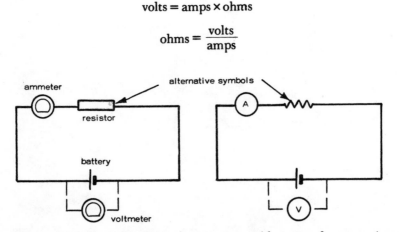

Fig. 7. Basic dc *circuit drawn in two ways, with meters for measuring current and voltage indicated*

This is one of the most basic and most useful laws of electronics and is equally applicable to *ac* circuits which are purely restive (i.e. do not have additional 'resistance' effects produced by the current being alternating rather than steady).

Ohm's Law makes it possible to calculate (and thus design) the performance of a simple *dc* circuit. For example, suppose we need a *current* of 200 milliamps (mA) to flow in a particular circuit to be powered by a 6 volt battery. Using Ohm's Law, the corresponding circuit resistance required to give this current can easily be worked out:

$$\text{ohms} = \frac{\text{volts}}{\text{amps}} = \frac{6}{0.200}$$

$$= 30 \text{ ohms.}$$

Components are connected by wires, but the resistance of wiring is small enough to be negligible. Thus in a simple *dc* circuit it is the effective total of all the resistor values and other components which offer resistance. Just what this total value is depends on how the various resistors which may be present are connected — *see* Chapter 4.

In some cases it is easy to calculate the resistance of a typical load. For example, a flashlight bulb is usually rated by volts and the current it draws. Ohm's law can then be used to find its nominal resistance. For example, if a bulb is rated at 6 volts and 50 mA, from Ohm's Law:

$$\text{Resistance} = \frac{6}{0.05} = 120 \text{ ohms.}$$

There is just one snag to this method of estimating load resistance. With filament bulbs, for example, the specified current drawn refers to the bulb in 'working' conditions with the filament heated up. Its actual resistance initially when the filament is cold can be considerably lower, drawing more current through the bulb. This may, or may not, be a disadvantage in a particular circuit. Also there are other types of load, like *dc* electric motors, where the effective resistance varies considerably with the speed at which the motor is running. Initially such a motor will have a very low resistance, its effective resistance then increasing with speed.

Two other basic relationships also apply in a simple *dc* circuit:

1. The *current* value will be the same through every part of the

circuit, unless a part of the circuit involves parallel-connected paths.

Thus in a circuit (A) of Fig. 8, all the resistors in the circuit are connected in *series*, so the *same* current will flow through each resistor.

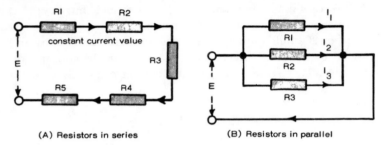

(A) Resistors in series   (B) Resistors in parallel

Fig. 8. Current has the same value through all resistors connected in series; but is different through each resistor connected in parallel

In circuit (B) of Fig. 8 the resistors are connected in *parallel*. In this case each resistor represents a separate path for the current and the value of current flowing through each 'leg' will depend on the value of that resistor. These current values can be calculated from Ohm's Law:

$$\text{through resistor 1, current} = \frac{E}{R1}$$

$$\text{through resistor 2, current} = \frac{E}{R2}$$

$$\text{through resistor 3, current} = \frac{E}{R3}$$

The current flowing through the 'wiring' part of the circuit will be the sum of these three currents,

$$\text{i.e. } \frac{E}{R1} + \frac{E}{R2} + \frac{E}{R3} \text{ or } E\left(\frac{1}{R1} + \frac{1}{R2} + \frac{1}{R3} \right)$$

2. The *voltage* throughout a simple *dc* circuit is not constant but will suffer a 'drop' across each resistor. This can be illustrated by the circuit shown in Fig. 9 where the voltages across the individual resistors are calculated (or measured with a voltmeter) as V1, V2 and V3.

The total resistance in the circuit is R1 + R2 + R3

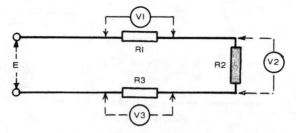

Fig. 9. Resistors 'drop' voltage in a dc circuit

The current (which will be the same throughout the circuit) is given by;

$$I = \frac{E}{R1 + R2 + R3}$$

We then have the conditions:

V1, measured across R1 = current × resistance
$$= 1 \times R1$$
V2, measured across R2 = 1 × R2
V3, measured across R3 = 1 × R3
Each of these voltages will be *less* than E.

Comparison with a hydraulic circuit again (*see* Chapter 2) can help understand how a resistor works as a 'voltage dropper'. In a hydraulic circuit, 'pressure' is analogous to 'voltage' in an electronic circuit. The equivalent to a resistor is some device restricting fluid flow — say a partially closed valve. Flow through this 'resistor' will produce a *pressure* drop. Similarly, the flow of electricity through a resistor will produce a *voltage* drop.

Voltage dropper circuit
The above is now reworked as a practical example. It is wanted to power a 6 volt electrical appliance (say a 6 volt transistor radio) from a 12 volt battery (e.g. a car battery). In this case the appliance is considered as a resistance *load*. To 'drop' the voltage from 12 to 6 across this load, a dropper resistor R is required in the circuit shown in Fig. 10. It remains to calculate a suitable value for this 'dropping' resistor, but to do this it is necessary to know the effective resistance of the load. (If this is not known it can be measured with an ohmmeter.) Suppose it is 100 ohms.

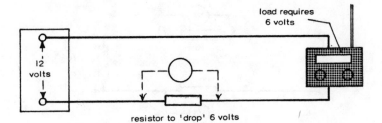

Fig. 10. Practical application of a dropping resistor

Using Ohm's Law again, if this load is to have 6 volts applied across it, and its resistance is 100 ohms, the current required to flow through the circuit is

$$1 = \frac{6}{100}$$
$$= 0.06 \text{ amps (60 milliamps)}$$

This same current will flow through the rest of the circuit. Thus considering the circuit from the 12 volt end:

$$\text{total resistance required} = \frac{12}{0.06}$$
$$= 200 \text{ ohms.}$$

The load already contributed 100 ohms, so the value of dropping resistor required must be $200-100 = 100$ ohms. A further calculation will show that the voltage drop across this resistor will be:

$$V = 0.06 \times 100$$
$$= 6 \text{ volts.}$$

This particular example also demonstrates another simple rule as regards dropping resistors. If the voltage is to be halved, then the value of the dropping resistor required is the same as that of the load.

Power in the Circuit
The power developed in a circuit by virtue of the electrical pressure (volts) and resulting current flow (amps) is given by the product of these two values, and measured in *watts*. Thus:

power = watts = volts × amps

This same definition applies both to *dc* and *ac* circuits.

Power is used up in producing a useful result in making the circuit 'work' (whether this be operating a radio, driving an electric motor, heating an electric element, etc). But all components which have resistance *absorb* a certain amount of power which is 'waste' power normally dissipated in the form of heat. No practical circuit can work without some resistance in circuit, and thus some power loss is inevitable. More important, the heating effect must not be so great that the component is damaged. Thus components normally have a *power rating* which should not be exceeded. In specific cases, even when operating within their power rating, provision may have to be made to conduct heat away from the component — as in the case of 'heat sinks' used with power transistors.

Referring to the example of the dropping resistor, this definitely wastes power to the tune of 6 (volts) × 0.06 (amps) = 0.36 watts. To be on the safe side, therefore, the resistor chosen would need to have a power rating of at least $\frac{1}{2}$ watt, and would also have to be placed in a position where it receives adequate ventilation to prevent heat build-up in the surrounding air.

The majority of transistor circuits work on low voltages, with low current values, and so components with quite moderate power ratings are usually adequate. Circuits carrying higher voltages and currents demand the use of components with correspondingly higher power ratings, and often need even more attention to ventilation. Thus the actual value of a component is only part of its specification. Its power rating can be equally important.

Note that since $V = 1R$ power can also be calculated as:

$$watts = (current)^2 \times Resistance$$
$$= 1^2R$$

This is often a more convenient formula for calculating power in a particular part of a circuit.

Shunt Circuits

A shunt circuit is used to 'drop' a *current* flowing through a particular component. It normally comprises two resistances in parallel, one resistance being the component resistance and the other the 'shunt' resistance. The appropriate value of the shunt resistance is again calculated directly from Ohm's Law.

A typical example of the use of a shunt resistance is to adapt an ammeter movement to measure different current ranges (as in a multimeter). In this case the 'load' resistance is that of the coil of the ammeter, which is initially designed to give a full scale deflection with a particular current flowing through it (call this I_1). The instrument cannot measure any higher current than I_1 since this would simply tend to carry the pointer past its full deflection, and very likely cause damage. Thus the meter is designed to handle the *lowest* current range required, and a shunt resistor (or a series of shunt resistors) added which can be switched into the meter circuit to extend the range. Fig. 11 shows this arrangement with just one shunt resistor connected for switching into the circuit.

If the shunt resistor is to extend the ammeter range to a higher current I_2 giving full scale deflection, then the required value of the shunt resistor follows from:

1. Current which has to flow through shunt is $I_2 - I_1$. This means that a current greater than I_1 will never flow through the meter movement (unless the actual current applied to the meter exceeds I_2).

2. Voltage drops across the meter = $I_1 \times R_m$ (where R_m is the resistance of the meter).

3. Shunt resistance required is therefore:

$$\frac{\text{voltage drop across instrument}}{\text{current flow through shunt}}$$

$$= \frac{I_1 \times R_m}{I_2 - I_1}$$

Again there is a simple rule to follow if the current range of the meter is to be *doubled*. In this case the shunt resistance required is the same as that of the meter.

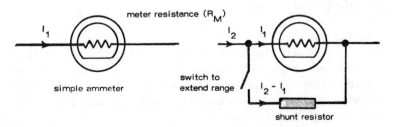

Fig. 11. Extending the range of a milliammeter

Ammeter into Voltmeter

An ammeter, which is an instrument used for measuring current, can also be made to measure volts by connecting a resistor in *series* with the meter — Fig. 12. This, in fact, is another example of a voltage dropper. Again, if the maximum meter current for full scale deflection is I_1, the *total* resistance which must be in circuit is:

$$\text{total } R = \frac{V}{I_1}$$

where V is the voltage range it is desired to measure.

The value of the series resistor required is this total resistance *less* the resistance of the meter (the latter may be negligible in comparison with the value of series resistor required and its likely tolerance — *see* Chapter 4).

Again several series resistors may be used, switched into the circuit individually to provide different voltage-measuring ranges on the meter movement, as shown in the right hand diagram of Fig. 12.

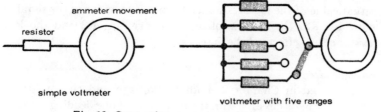

Fig. 12. Converting an ammeter into a voltmeter

Potential Dividers

A potential divider is yet another example of the practical application of a 'voltage dropper'. The basic circuit is shown in Fig. 13. and since the *current* flow through R1 and R2 is the same, the following voltage values apply:

V1 = source voltage (e.g. battery voltage)

$$V2 = \frac{V1}{R1 + R2} \times R1$$

$$V3 = \frac{V1}{R1 + R2} \times R2$$

(Note: $\dfrac{V1}{R1 + R2}$ is the current flowing through R1 and R2).

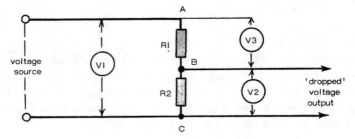

Fig. 13. Basic potential divider circuit

It follows that by suitable selection of values for R1 and R2, virtually any lower voltage than V1 can be tapped from points A and B, or B and C (or both). It also has the advantage that it is not necessary to know the load resistance before suitable 'dropper' resistances can be calculated. It could thus be a more practical alternative for the example described in Fig. 10, but connection to a *load* will, of course, result in a further drop in voltage.

If the resistance of the load is known, then there is no particular problem with a fixed resistor potential divider. Calculate the value of R2 (Fig. 13) on the basis of no load resistance, then subtract the actual value of the load resistance from this to arrive at the required value for R2. (In the complete tapped circuit, R2 and the load resistance will effectively be in series.)

Basic ac circuits
As explained in Chapter 2, both the voltage and current flow is alternating in *ac* circuits, with the possibility of one 'leading' or 'lagging' the other. Also it was intimated that the effective resistance offered by resistance components may be modified (usually increased) by reactive effects. These effects become increasingly marked as the *frequency* of the *ac* increases, and at radio frequencies are more pronounced than pure resistance.

It is possible to obtain an *ac* circuit which is purely resistive, particularly at lower frequencies, in which case Ohm's Law is equally valid for such circuits as it is for *dc* circuits. Ohm's Law can also be applied to *ac* circuits in which reactive effects *are* present, but in a slightly modified form. These reactive effects are described specifically as *reactance* and *impedance*.

Reactance is the circuit loading effect produced by *capacitors* and *inductances* (coils). It is measured in ohms and designated by the symbol X. Its actual value is dependent both on the component value and the frequency of the *ac*.

In the case of capacitors, capacitive reactance (usually designated X_c) is given by:

$$(X_c)_c = \frac{1}{2\pi f C}$$

where f is the *ac* frequency in Hz

C is the capacitance in farads

$\pi = 3.1412$

In the case of inductances, *inductive reactance* (usually designated X_L) is given by:

$$X_L = 2\pi f L$$

where L is the inductance in Henrys

If the *ac* circuit contains only reactance (i.e. does not have any separate resistance), then X takes the place of resistance (R) in the Ohm's Law formula:

$$I = \frac{E}{X}$$

In practice, reactance present is also usually associated with resistance, the resulting combination representing the *impedance* (Z) of the circuit.

If reactance and resistance are in series:

$$Z = \sqrt{R^2 + X^2}$$

If reactance and resistance are in parallel:

$$Z = \sqrt{\frac{RX}{R^2 + X^2}}$$

And again impedance (Z) takes the place of resistance in the Ohm's Law formula:

$$I = \frac{E}{Z}$$

These are then the basic formulas for *ac* circuit calculations.

Power Factor

Power factor is something specific to *ac* circuits, although it is only the *resistance* in such circuits that actually consume power. This power consumed can be calculated as the product of the square of the current flowing through the resistance and the value of the resistance, i.e. I^2R watts. The *apparent* power in the circuit is the product of *ac* voltage and current, correctly specified as volt-amps.

The ratio of the power consumed to the apparent power is called the *power factor*, usually expressed as a percentage. If the circuit is purely resistive, then the power factor will be 100 per cent (since all the apparent power is consumed in the resistance). Reactance does not consume power, so in a purely reactive circuit the power factor is zero. When a circuit contains both resistance and impedance (i.e. reactance), then the power factor will always be less than 100 per cent, its value depending on the resistance present.

If it seems strange that reactance does not consume power, the explanation is quite simple. With alternating voltage and current, any energy absorbed by a capacitor or inductance on one half of the cycle is returned on the other half of the cycle.

It should also be made clear that 'power consumed' by a resistance in an *ac* circuit is not a power *loss*. It is simply power which is transferred backwards and forwards without any heating effect.

dc and ac in the same circuit

It is quite possible to have both *dc* and *ac* flowing in the same circuit. In fact, this is the principle on which most radio and similar circuits work. The *dc* is the basic source of electrical supply, on which various *ac* currents are superimposed. The one essential difference is that *dc* can only flow through a continuous circuit, whereas *ac* can pass through components such as capacitors which present a 'break' in the circuit to *dc*. These effects can be used to advantage to 'isolate' stages in a circuit.

In the type of circuit shown in Fig. 14, for example, an input comprising a mixture of *dc* and *ac* is applied to the left hand side of the circuit. If only the *ac* component of the signal is required, the *dc* content can be blocked by a capacitor (C1). Meantime the next part of the circuit or 'stage' which has to deal with that signal is 'powered' by *dc* from the source supply (say a battery), probably via resistors R1 and R2 acting as potential dividers to get the voltages correct for that stage (other stages may need different working *dc* voltages, all coming from the same source). Output signal from this stage then consists of a mixture of *dc* and *ac*. If only the *ac* content is wanted for passing to the next stage, a capacitor (C2) is again used as a block for *dc*.

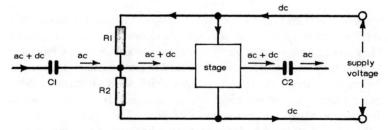

Fig. 14. Diagrammatic representation of flow of ac *and* dc *in a circuit*

CHAPTER 4

Resistors

Resistors, as their name implies, are designed to provide some desirable, or necessary, amount of *resistance* to current flow in a curcuit. They can also be used to 'drop' voltages, as explained in Chapter 3. As such they are the main elements used in circuit design to arrive at the desired current flows and voltages which work the circuit. Resistors do not generate electrical energy, but merely absorb it, which the resistor dissipates in the form of heat. The performance of a resistor is not affected by frequency, so it behaves in the same way in both *dc* and *ac* circuits. (There are exceptions, as noted later.)

Resistors are specified by (i) resistance value in ohms; (ii) tolerance as a percentage of the nominal value; and (iii) power rating in watts. They are also categorized by the type of construction.

Resistance value and tolerance is normally indicated by a colour code consisting of four coloured rings, starting at or close to one end — Fig. 15. These are read as follows:

1st ring gives first digit
2nd ring gives second digit
3rd ring gives number of noughts to put after first two digits.

The universally adopted colour coding is:

Colour	1 gives first figure of resistance value	2 gives second figure of resistance value	3 gives number of noughts to put after first two figures
Black	0	0	None
Brown	1	1	0
Red	2	2	00
Orange	3	3	000
Yellow	4	4	0000
Green	5	5	00000
Blue	6	6	000000
Violet	7	7	0000000
Grey	8	8	00000000
White	9	9	000000000

Example: resistor colour code read as Brown, Blue, Orange.

	Brown	Blue	Orange
value read as	1	6	000

i.e. 16,000 Ω or 16 kΩ (kilohms).

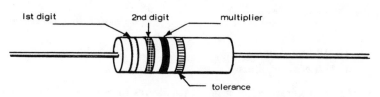

Fig. 15. Standard resistor colour code marking

The fourth coloured ring gives the tolerance, viz:

    silver — 10% tolerance either side of the nominal value
    gold — 5% tolerance either side of the nominal value
    red — 2% tolerance either side of the nominal value
    brown — 1% tolerance either side of the nominal value

Absence of a fourth ring implies a tolerance of 20%.

Certain types of modern resistors of larger physical size may have letters and numbers marked on the body instead of coloured rings. With this coding the numbers indicate the numerical value and the following letter the multiplier, where:

    E = ×1
    K = ×1000 (or kilohms)
    M = ×1000000 (or megohms)

A second letter then gives the tolerance:

    M = 20% tolerance either side of the nominal value
    K = 10% tolerance either side of the nominal value
    J = 5% tolerance either side of the nominal value
    H =2.5% tolerance either side of the nominal value
    G = 2% tolerance either side of the nominal value
    F = 1% tolerance either side of the nominal value

The actual range of (nominal) resistance values to which resistors are made is based on steps which give an approximately constant *percentage* change in resistance from one value to the next — not simple arithmetical steps like 1, 2, 3, etc. These are based on the *preferred* numbers:

1, 1.2, 1.5, 1.8, 2.2, 2.7, 3.3, 3.9, 4.7, 5.6, 6.8, 8.2, 10, 12, 15, 18, etc.

Thus, for example, a typical range of resistor values would be:

    10, 12, 15, 18, 22, 27, 33, 39, 47, 56, 68, 82, and 100 ohms
    120, 150, 180, 220, 270, 330, 390, 470, 560, 680 and 820 ohms
    1, 1.2, 1.5, 1.8, 2.2, 2.7, 3.3, 3.9, 4.7, 5.6, 6.8, and 8.2
    kilohms

10, 12, etc kilohms
1, 1.2, etc megohms

As regards tolerances, as a general rule resistors with a 10 per cent tolerance are suitable for average circuit use. The actual resistance value of, say, a 1 kohm resistor would then be anything between 900 and 1100 ohms. For more critical work, such as radio circuits, resistors with a 5 per cent tolerance would be preferred. Closer tolerances are not normally required, except for very critical circuits.

Power Rating
The physical size (or shape) of a resistor is no clue to its resistance value, but can be a rough guide to its *power rating*. Physical sizes (Fig. 16) range from about 4mm long by 1mm diameter up to about 50mm long and 6mm or more diameter. The former would probably have a power rating of 1/20th watt and the latter possibly 10 watts. More specifically, however, the power rating is related to *type* as well as size. A general rule which does apply to power rating, however, is that whilst this figure nominally represents a safe maximum which the resistor can tolerate without damage, it is usually best to operate a resistor well below its power rating — say at 50 per cent rating — particularly if components are crowded on a circuit, or the circuit is enclosed in a case with little or no ventilation.

Voltage Rating
Maximum operating voltage may also be specified for resistors, but since this is usually of the order of 250 volts or more, this parameter is not important when choosing resistors for battery circuits. Resistors used on mains circuits must, however, have a suitable voltage rating.

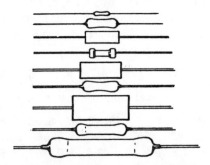

Fig. 16. Examples of modern resistor outlines (actual size)

Types of Construction

Resistor *types* classified by construction are:

1. *Carbon* (also called carbon-composition, moulded-carbon and carbon rod). These are in the form of a small rod moulded from carbon and a binder, with wire connections at each end. The rod is usually protected with a paper or ceramic sleeve, or a lacquer coating. These are the most common (and cheapest) type of resistor, generally available in values from 10 ohms to 22 megohms. Standard types are usually available in $\frac{1}{8}$, $\frac{1}{4}$, $\frac{1}{2}$, 1 and 2 watt ratings.

It is a general characteristic of carbon resistors that their value remains stable at normal temperatures, but above 60°C their resistance increases rapidly with increasing temperature.

2. *Carbon-film resistors* (also known as high-stability carbon resistors). These are manufactured by depositing a thin film of carbon on a small ceramic rod. The rod is fitted with metal end caps, to which wire leads are attached. The body of the resistor is usually protected by a varnish, paint or silicone resin coating, but some types may be encased in a ceramic, plastic or glass outer coating.

Carbon-film resistors are little affected by temperature changes (their stability is usually better than 1 per cent) and are also characterized by low 'noise'. They are available in sub-miniature sizes (1/20 and 1/10 watt power rating); and in larger sizes up to 1 watt power rating. They are a preferred type for radio circuits, particularly as they have excellent high frequency characteristics.

3. *Metal-film resistors*. These are made by depositing a metallic film (usually nickel-chromium) on a glass or ceramic rod. A helical track is then cut in the film to produce the required resistance value. Metallic end caps are then fitted, carrying the wire leads, and the body protected by a lacquer, paint or plastic coating. Stability characteristics are similar to those of carbon-film resistors, but they are more expensive. They are generally produced in miniature sizes with power ratings from 1/10 watt upwards.

4. *Metal-oxide film resistors*. Construction is similar to that of a metal-film resistor except that the coating used is a metallic oxide (usually tin oxide), subsequently covered with a heat-resistant coating. This type of resistor is virtually proof against accidental overheating (e.g. when making soldered connections) and is also not affected by damp. Stability is very high (better than 1 per cent), and the power ratings are high for their physical size.

5. *Metal-glaze resistors*. In this type the resistive film deposited on

the rod is a cermet (metal-ceramic); otherwise construction is similar
to metal-film resistors.

Film-resistors may also be classified as *thick-film* or *thin-film*. As a
general rule, individual resistors of this type are thick-film. Thick-
film resistors are also made in groups on a small substrate and en-
capsulated in integrated circuit 'chips'. Thin-film resistors are made
in a similar way, but on a considerably smaller scale for use in the
manufacture of integrated circuits.

Effect of Age
All resistors can be expected to undergo a change in resistance with
age. This is most marked in the case of carbon-composition resistors
where the change may be as much as 20 per cent in a year or so's use.
In the case of carbon-film and metallic-film resistors, the change is
seldom likely to be more than a few per cent.

Effect of High Frequencies
The general effect of increasing frequency in *ac* circuits is to decrease
the apparent value of the resistor, and the higher the resistor value
the greater this change is likely to be. This effect is most marked with
carbon-composition and wire-wound resistors (*see* below). Carbon-
film and metal-film resistors all have stable high frequency
characteristics.

Wire-wound resistors
Wire-wound resistors are made by wrapping a length of resistance
wire around a ceramic coil. The whole is then covered with a protect-
ive coating or film. The specific advantages offered by wire resistors
are that a wide range of values can be produced (typically from 1
ohm to 300 kilohms) with power ratings from 1 to 50 watts (or up to
225 watts in 'power' types), with tolerances as close as 1 per cent.
They also have excellent stability and low 'noise'. Their disadvantages are that they are more costly and also unsuitable for use in *ac*
circuits carrying high frequencies because their effective value
changes. Physically they need be no bigger than film-type resistors
for the same power rating.

Variable Resistors
Variable resistors consist of a resistive track swept by a wiper arm.
The position of the wiper arm determines the length of track in
circuit, and thus the actual resistance present. The track may be

circular (usually a 270° arc) or in a straight line, circular types being the more common. Both types are known as potentiometers or pots.

The resistive element may be wire-wound, carbon-composition, carbon-film or metallic-film. The former type is known as a wire wound potentiometer. Carbon-track potentiometers are the cheapest (with the same limitations as carbon-composition resistors), but are available only with moderate power ratings, e.g. ¼ watt for low resistance values, reducing with higher resistance values. Wire-wound potentiometers usually have higher power ratings and are also available in lower resistance values than carbon-track potentiometers. Tolerances are usually of the order of 10 per cent or 20 per cent, but may be much closer with *precision potentiometers.*

Connections should be obvious from Fig. 17. Thus with connections to end 1 of the track and the wiper, length 1 to C of the resistive track is in circuit. Actual resistance in circuit can thus be varied by moving the wiper towards 3 (increasing resistance in circuit); or towards 1 (decreasing resistance in circuit).

The change in resistance can occur *proportionally* to the actual length of track involved: or *logarithmetically*, where there is a logarithmic increase in resistance with wiper movement uncovering more track (similar to the 'steps' adopted for standard resistor values). The former is known as a *linear* potentiometer and the latter a *log* potentiometer. Potentiometers can also have characteristics between the two. Note that 'linear' in this description has quite a different meaning to a linear physical *shape* of potentiometer. To avoid confusion it is best to refer to the latter as a *slide-type* potentiometer.

There is also a class of variable resistors intended to be adjusted to a particular resistance setting and then left undisturbed. These are known as *preset potentiometers* (alternatively, *preset pots* or just

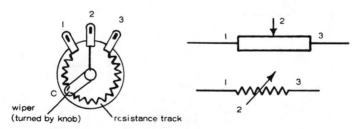

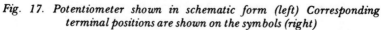

Fig. 17. Potentiometer shown in schematic form (left) Corresponding terminal positions are shown on the symbols (right)

presets). They are small in size and more limited in maximum resistance value — typically from 100 ohms to 1 megohm. They are usually designed for adjustment by a screwdriver applied to the central screw, or sometimes by a knurled disc attached to the central spindle, carrying the wiper. The latter type are known as *edge presets* and similar types, with a (maximum) resistance of 5 k ohms may be used as volume controls on miniature transistor radios.

Potentiometers are used specifically in a circuit where it is necessary to be able to adjust resistance. A typical example is the volume control in a radio circuit. In this case the potentiometer may be designed so that at one end of the track the wiper runs right off the track to break the circuit. Thus the volume control can also be connected up to work as an on-off switch, using this extra facility provided.

Another practical example is the replacement of fixed resistors in a potential divider by a single potentiometer to make the circuit variable in performance. Thus the circuit previously described in Fig. 13 (chapter 3) always gives a predetermined voltage at the tapping points (provided the supply voltage remains constant). Replacing resistors R1 and R2 with a potentiometer, as shown in Fig. 18, with the tapping point taken from one end of the potentiometer, and the wiper will give a tapped voltage which is fully variable from the full supply voltage down to zero, depending on the position of the wiper.

In practice in a variable voltage circuit of this type it may be necessary to leave a fixed resistor in series with the potentiometer to limit the current being drawn in the event that the potentiometer has been adjusted to zero resistance and the tapped circuit is broken or switched off with the original supply still connected. Without the fixed resistor, the supply would then be shorted. The value of a fixed resistor would be calculated to limit the current drawn in such a case to a safe level.

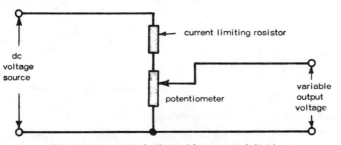

Fig. 18. A practical adjustable potential divider

With a fixed resistor in series with the potentiometer, of course, the maximum voltage that can be 'tapped' from the potentiometer is equal to the supply voltage less the voltage dropped by the fixed resistor.

The main thing to watch in such a circuit is that the power rating of the potentiometer is adequate to accommodate the voltage and current drain in the tapped circuit. But it has one further advantage over a fixed resistor potential divider. When a load is added to the tapped circuit, this will add resistance in that circuit, causing a further voltage drop. Unless this is allowed for in calculating the values for the fixed resistors in a potential divider, this will mean that the load receives *less* than the design voltage. With a potentiometer replacing the two fixed resistors, its position can be adjusted to bring the load voltage back to the required figure, Fig. 19. This considerably simplifies the design of a potential divider where the load resistance is known only approximately, or not at all.

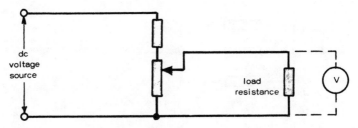

Fig. 19. The potentiometer can be adjusted to give required voltage across the load

Circuit Rules for Resistors

In the case of resistances connected in *series* (Fig. 20), the total resistance in circuit will be the sum of the various resistor values, i.e.

$$\text{total resistance} = R1 + R2 + R3 + \ldots \ldots$$

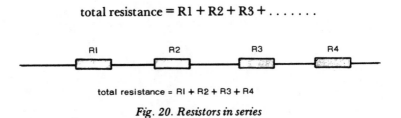

total resistance = RI + R2 + R3 + R4

Fig. 20. Resistors in series

In the case of resistors connected *in parallel* (Fig. 21), the total effective resistance is given by:

$$\frac{1}{R} = \frac{1}{R1} + \frac{1}{R2} + \frac{1}{R3} + \ldots \ldots$$

where R is the total resistance

In the case of two dissimilar resistors

$$R = \frac{R1.R2}{R1 + R2}$$

or remembered as

$$\text{total resistance} = \frac{\text{product of resistor values}}{\text{sum of resistor values}}$$

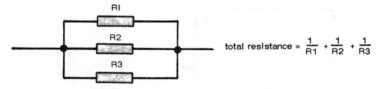

Fig. 21. Resistors in parallel

CHAPTER 5

Capacitors

A capacitor is basically a device which stores an electric charge. Physically it consists of two metal plates or electrodes separated by an insulating material or *dielectric*. Application of a *dc* voltage across the capacitor will produce a deficiency of electrons on the positive plate and excess of electrons on the negative plate — Fig. 22. This differential accumulation of electrons represents an electric charge, which builds up to a certain level (depending on the voltage) and then remains at that level.

As far as *dc* is concerned, the insulator acts as a *blocking* device for current flow (although there will be a certain transient charging current which stops as soon as the capacitor is fully charged). In the case of *ac* being applied to the capacitor the charge built up during one half cycle becomes reversed on the second half of the cycle, so that effectively the capacitor conducts current through it as if the dielectric did not exist. Thus as far as *ac* is concerned, a capacitor is a *coupling* device.

There are scarcely any electronic circuits carrying *ac* which do not incorporate one or more capacitors, either for coupling or shaping the overall *frequency response* of the network. In the latter case a capacitor is associated with a resistor to form an *RC combination* (*see* Chapter 6). The charge/discharge phenomenon associated with capacitors may also be used in other types of circuits (e.g. the photographic electronic flash is operated by the charge and subsequent discharge of a capacitor triggered at the appropriate moment).

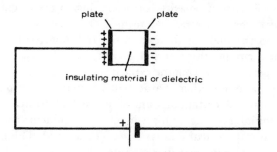

Fig. 22. Illustrating how a capacitor builds up a charge when connected to a dc voltage, blocking current flow

Like resistors, capacitors may be designed to have fixed values or be variable in capacity. Fixed capacitors are the main building blocks of a circuit (together with resistors). Variable capacitors are mainly used for adjusting tuned circuits.

Fixed capacitors fall into two main categories:
(i) non-polarized capacitors and (ii) polarized or *electrolytic* capacitors. The main thing which determines the type of capacitor is the dielectric material used.

Non-polarized capacitors consist, basically, of metallic foil inter-leaves with sheets of solid dielectric material, or equivalent construc-tion. The important thing is that the dielectric is 'ready made' before assembly. As a consequence it does not matter which plate is made positive or negative. The capacitor will work in just the same way, whichever way round it is connected in a circuit, hence the descrip-tion 'non-polarized'. This is obviously convenient, but this form of construction does limit the amount of capacitance which can be accommodated in a single 'package' of reasonable physical size. Up to about 0.1 microfarads, the 'package' can be made quite small, but for capacitance values much above 1 microfarad, the physical size of a non-polarized capacitor tends to become excessively large in com-parison with other components likely to be used in the same circuit.

This limitation does not apply in the case of an *electrolytic* capacitor. Here initial construction consists of two electrodes separ-ated by a thin film of *electrolyte*. As a final stage of manufacture a voltage is applied across the electrodes which has the effect of pro-ducing a very thin film of non-conducting metallic oxide on the surface of one plate to form the dielectric. The fact that capacitance of a capacitor increases the thinner the dielectric is made means that very much higher capacities can be produced in smaller physical sizes. The only disadvantage is that an electrolytic capacitor made in this way will have a polarity corresponding to the original polarity with which the dielectric was formed, this correct polarity being marked on the body of the capacitor. If connected the other way round in a circuit, the reversed polarity can destroy the dielectric film and permanently ruin the capacitor.

There is also one other characteristic which applies to an electro-lytic capacitor. A certain amount of 'unused' electrolyte will remain after its initial 'forming'. This will act as a conductor and can make the capacitor quite 'leaky' as far as *dc* is concerned. This may or may not be acceptable in particular circuits.

Non-polarized Capacitor Types

Various types of construction are used for non-polarized capacitors, most of which are easily identified by the shape of the capacitor — *see* Fig. 23. There is no need to go into detail about the actual constructions. Their specific characteristics are important, though, as these can determine the best type to use for a particular application.

1. *Paper dielectric capacitors*, generally recognizable by their tubular form, are the least expensive but generally bulky, value for value, compared with more modern types. Their other main limitation is that they are not suitable for use at frequencies much above 1 MHz, which virtually restricts their application to *af* circuits. They are generally available in capacities from 0.05 μF up to 1 or 2 μF, with working voltages from 200 to 1000 volts. Plastic-impregnated paper dielectric capacitors may have much higher working voltages.

2. *Ceramic capacitors* are now widely used in miniaturized *af* and *rf* circuits. They are relatively inexpensive and are available in a wide range of capacities from 1 pF to 1 μF with high working voltages and also characterized by high leakage resistance. They are produced in both disc and tubular shapes; also as metallized ceramic plates.

3. *Silver-mica capacitors* are more expensive than ceramic capacitors but have excellent high frequency response and much smaller tolerances, so are generally regarded as superior for critical *rf* applications. They can be made with very high working voltages.

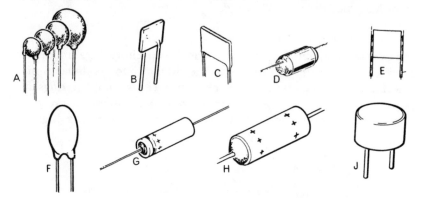

Fig. 23. Examples of modern capacitors

A — *ceramic disc*	F — *tantalum bead*
B — *ceramic plate*	G — *electrolytic (polarized)*
C — *silvered mica*	H — *non-polarized electrolytic*
D — *polystyrene*	J — *plug-in type (Siemens)*
E — *polycarbonate*	

4. *Polystyrene capacitors* are made from metallic foil interleaved with polystyrene film, usually with a fused polystyrene enclosure to ensure high insulation resistance. They are noted for their low losses at high frequencies (i.e. low inductance and low series resistance). good stability and reliability. Values may range from 10 pF to 100,000 pF, but working voltage generally falls substantially with increasing capacity (e.g. as low as 60 volts for a 100,000 pF polystyrene capacitor).

5. *Polycarbonate capacitors* are usually produced in the form of rectangular slabs with wire end connections designed to plug into a printed circuit board. They offer high values of capacity (up to 1 μF) in very small sizes, with the characteristics of low losses and low inductance. Like polystyrene capacitors, working voltages become more restricted with increasing capacity value.

6. *Polyester film capacitors* are also designed for use with printed circuit boards, with values from 0.01 μF up to 2.2 μF. Value for value they are generally larger in physical size than polycarbonate capacitors. Their low inherent inductance makes them particularly suitable for coupling and decoupling applications.

Values of polyester film capacitors are indicated by a colour code consisting of five colour bands, reading from the top — *see* Fig. 24.

7. *Mylar film capacitors* can be regarded as a general purpose film type, usually available in values from 0.001 μF up to 0.22 μF with a working voltage up to 100 volts *dc*.

Electrolytic Capacitors
The original material used for electrolytic capacitors was aluminium foil, together with a paste electrolyte, wound into a tubular form with an aluminium outer cover, characterized by 'dimpled' rings at one or both ends. The modern form of aluminium electrolytic capacitor is based on etched foil construction, enabling higher

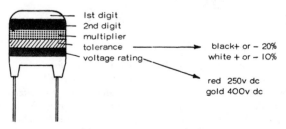

Fig. 24. Colour code for polyester capacitors

capacitance values to be achieved in smaller can sizes. Values available range from 1 μF up to 4700 μF (or even larger, if required). Working voltages are generally low, but may range from 10 volts *dc* up to 250 or 500 volts *dc*, depending on value and construction. A single lead emerges from each end, but single-ended types are also available (both leads emerging from one end); and can-types with rigid leads in one end for plugging into a socket. Single-ended types are preferred for mounting on printed circuit boards.

The other main type of electrolytic is the *tantalum capacitor*. This is produced both in cylindrical configuration with axial leads, or in *tantalum bead* configuration. Both (and the latter type particularly) can offer very high capacitance values in small physical sizes, within the range 0.1 to 100 μF. Voltage ratings are generally low, e.g. from 35 volts down to less than 10 volts *dc*.

All electrolytic capacitors normally have their value marked on the body or case, together with a polarity marking (+ indicating the positive lead). Tantalum bead capacitors, however, are sometimes colour coded instead of marked with values. This colour coding is shown in Fig. 25, while other codes which may be found on other types of non-polarized capacitors are given in Fig. 26.

Tolerance of Fixed Capacitors

As a general rule, only silver-mica capacitors are made to close tolerances (plus or minus 1 per cent is usual). The tolerance on other

	Capacitance in μF			3rd Ring *dc*	Working Voltage
COLOUR	1st Ring	2nd Ring	Spot Polarity and Multiplier	Colour	Voltage
Black	—	0	× 1.00	White	3
Brown	1	1	× 10	Yellow	6.3
Red	2	2		Black	10
Orange	3	3		Green	16
Yellow	4	4		Grey	25
Green	5	5		Pink	35
Blue	6	6			
Violet	7	7			
Grey	8	8	× 0.01		
White	9	9	× 0.10		

Fig. 25. Colour code for tantalum bead capacitors

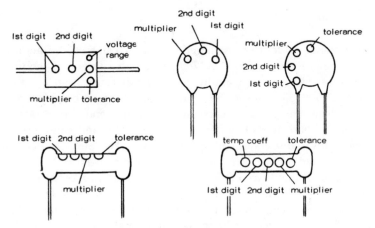

Fig. 26. Other coding systems used on capacitors

types of capacitors is usually between 10 and 20 per cent and may be even higher (as much as 50 per cent) in the case of aluminium foil electrolytics. Because of the wide tolerances normal with electrolytics, choice of actual value is seldom critical.

Variable Capacitors

Variable capacitors are based on interleaved sets of metal plates, one set being fixed and the other movable. The plates are separated by a dielectric which may be air or a solid dielectric. Movement of one set of plates alters the effective area of the plates, and thus the value of capacitance present.

There is also a general distinction between *tuning capacitors* used for frequent adjustment (e.g. to tune a radio receiver to a particular station) and *trimmer capacitors* used for initial adjustment of a tuned circuit. Tuning capacitors are larger, more robust in construction and generally of air-dielectric type. Trimmer capacitors are usually based on a mica or film dielectric with a smaller number of plates, capacity being adjusted by turning a central screw to vary the pressure between plates and mica. Because they *are* smaller in size, however, a trimmer capacitor may sometimes be used as a tuning capacitor on a sub-miniature radio circuit, although special miniature tuning capacitors are made for radios designed to mount directly on a printed circuit board.

In the case of tuning capacitors the shape of the vanes determines the manner in which capacitance changes with spindle movement. These characteristics usually fall under one of the following descriptions:

1. *Linear* — where each degree of spindle rotation produces an equal change in capacitance. This is the most usual type chosen for radio receivers.

2. *Logarithmic* — where each degree of spindle movement produces a constant *percentage* change in *frequency* of a tuned circuit.

3. *Even frequency* — where each degree of spindle movement produces an *equal* change in frequency in a tuned circuit.

4. *Square law* — where the change in capacitance is proportional to the *square* of the angle of spindle movement.

Basic Circuit Rules for Capacitors

The rules for total capacitance of capacitors in series and in parallel is the opposite way round to that for resistors. For capacitors connected in *series* (Fig. 27), the total effective capacity (C) is given by:

$$\frac{I}{C} = \frac{I}{C1} + \frac{I}{C2} + \frac{I}{C3} + \ldots$$

or in the case of two dissimilar capacitors

$$C = \frac{C1.C2}{C1 + C2}$$

In words,

$$\text{total capacitance} = \frac{\text{product of capacitances}}{\text{sum of capacitances}}$$

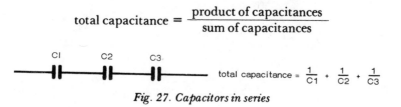

total capacitance $= \frac{1}{C1} + \frac{1}{C2} + \frac{1}{C3}$

Fig. 27. Capacitors in series

For capacitors connected in parallel (Fig. 28)

$$C = C1 + C2 + C3 + \ldots$$

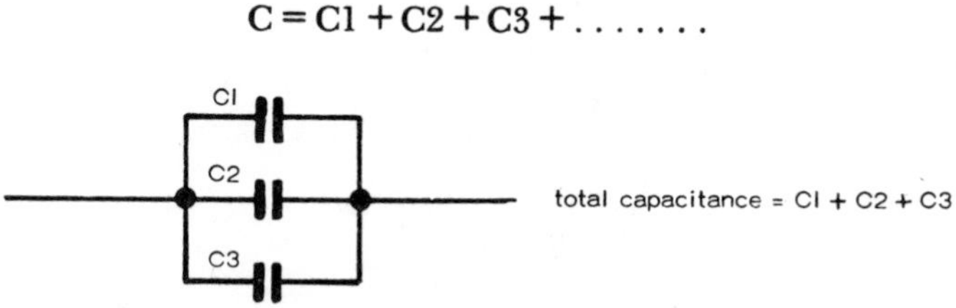

total capacitance $= C1 + C2 + C3$

Fig. 28. Capacitors in parallel

This capacitance effect, of course, is only apparent in an *ac* circuit. In a *dc* circuit a capacitor simply builds up a charge without passing current. In a practical *ac* circuit a capacitor also exhibits *reactance* (*see* Chapter 2), and because of its construction may also exhibit a certain amount of *inductance* (*see* Chapter 7).

CHAPTER 6

Capacitor and RC Circuits

One of the principal uses of a capacitor is as a coupling device capable of passing *ac* but acting as a block to *dc*. In any practical circuit there will be some resistance connected in series with the capacitor (e.g. the resistive load of the circuit being coupled). This resistance limits the current flow and leads to a certain delay between the application of a voltage to the capacitor and the build-up of charge on the capacitor equivalent to that voltage. It is this 'charge voltage' which blocks the passage of *dc*. At the same time the combination of resistance with capacitance, generally abbreviated to RC, will act as a *filter* capable of passing *ac* frequencies, depending on the charge-discharge time of the capacitor, or the *time constant* of the RC combination.

The formula for calculating the time constant (T) is quite simple:

$$T = RC$$
where T = time constant in seconds
R = resistance in megohms
C = capacitance in microfarads

(It can be noted that the same numerical value for T is given if R is in ohms and C in farads, but megohms and microfarads are usually much more convenient units).

The *time constant* is actually the time for the voltage across the capacitor in an RC combination to reach 63 per cent of the applied voltage (this 63 per cent figure being chosen as a 'mathematical convenience'). The voltage across the capacitor will go on building up to almost (but never quite) 100 per cent of the applied voltage, as shown in Fig. 29.

The time constant factor refers to the duration of time in terms of the time factor, e.g. at 1 (which *is* the time factor of the RC combination) 63 per cent full voltage has been built up, in a time equal to 2 × the time constant, 80 per cent full voltage; and so on. After a time constant of 5 the full (almost 100 per cent) voltage will have been built up across the capacitor.

The discharge characteristics of a capacitor take place in essentially the inverse manner, e.g. after a period of time equal to the time

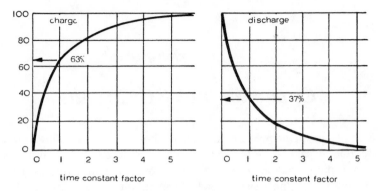

Fig. 29. Percentage voltage across capacitor related to time when being charged (left) and discharged (right)

constant the voltage across the capacitor will have dropped to $100 - 63 = 37$ per cent of the full voltage; and so on.

In theory, at least, a capacitor will never charge up to full applied voltage; nor will it fully discharge. In practice, full charge, or complete discharge, can be considered as being achieved in a period of time equal to five time constants. Thus in the circuit identified with Fig. 30, closing switch 1 will produce a 'full' charge on the capacitor in 5 × time constant seconds. If switch 1 is now opened, the capacitor will then remain in a condition of storing a voltage equivalent to the original applied voltage, holding this charge indefinitely if there is no internal leakage. In practice it will very slowly lose its charge as no practical capacitor is perfect, but for some considerable time it will remain effectively as a potential source of 'full charge' voltage. If the capacitor is part of a mains circuit, for example, it is readily capable

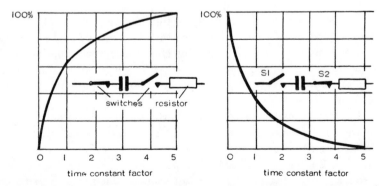

Fig. 30. Closing switch 1 allows capacitor to charge right up in a time equal to 5 time constants. It then takes a similar period to discharge fully through a load resistor when switch 2 is closed

of giving an 'electric shock' at mains voltage if touched for some time after the circuit has been switched off.

To complete the cycle of charge-discharge as shown in the second diagram of Fig. 30 Switch 2 is closed, when the capacitor discharges through the associated resistance, taking a finite amount of time to complete its discharge.

Fig. 31 shows a very simple circuit working on this principle. It consists of a resistor (R) and capacitor (C) connected in series to a source of *dc* voltage. As a visual indication of the working of the circuit, a neon is connected in parallel with the capacitor. A neon virtually represents an open circuit until its threshold voltage is applied, when it immediately conducts current like a low resistance and glows (*see* Chapter 10 for more about neons). The voltage source for this circuit must therefore be above that of the neon turn-on voltage.

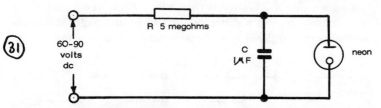

Fig. 31. Simple neon flasher circuit (note the symbol for a neon). The values of the resistor (R) and capacitor (C) determine the flashing rate

When this circuit is switched on, the capacitor starts to build up a charge at a rate depending on the time constant of R and C. The neon is 'fed' by voltage developed across the capacitor. Once this reaches the turn-on voltage of the neon, the neon will switch on and cause the capacitor to discharge through the neon causing it to light. Once the capacitor has discharged, no more current flows through the neon and so it switches off again until the capacitor has built up another charge equivalent to the turn-on voltage, when it will discharge through the neon; and so on. In other words the neon will flash at a rate determined by the time constant of R and C.

Using the component values shown, the time constant for the circuit is:

$$T = 5 \text{ (megohms)} \times 0.1 \text{ (microfarads)}$$
$$= 0.5 \text{ seconds.}$$

This is not necessarily the actual *flashing rate* of the circuit. It may

take a period of more than one time constant (or less) for the
capacitor voltage to build up to the neon turn-on voltage — more if
the turn-on voltage is greater than 63 per cent of the supply voltage;
less if the turn-on voltage is less than 63 per cent of the supply
voltage.

It also follows that the flashing rate can be altered by altering the
value of R or C, either by substituting different values calculated to
give a different time constant; or with a parallel connected resistor or
capacitor. Connecting a similar value resistor in parallel with R, for
example, would *halve* the flashing rate (since paralleling similar
resistor values halves the total resistance). Connecting a similar value
capacitor in parallel with C would *double* the flashing rate.

This type of circuit is known as a *relaxation oscillator*. Using a
variable resistor for R it could be adjusted for a specific flashing rate.
It can also be extended in the form of a novelty lighting system by
connecting a series of RC circuits each with a neon in cascade, each
RC combination having a different time constant — Fig. 32. This
will produce random flashing of the neons in the complete circuit.

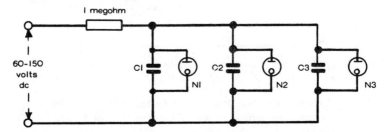

*Fig. 32. Random 'cascade' flasher circuit. Any number of neons can be
connected in this way and will flash in random order*

Capacitors in ac Circuits

As far as *ac* is concerned, the fact that the applied voltage is alternating means that during one half cycle the capacitor is effectively
being charged and discharged with one direction of voltage; and
during the second half of the *ac* cycle, charged and discharged with
opposite direction of voltage. Thus, in effect, *ac* voltages pass
through the capacitor, restricted only by such limitations as may be
applied by the RC *time constant* which determines what *proportion*
of the applied voltage is built up and discharged through the
capacitor. At the same time a capacitor will offer a certain opposition to the passage of *ac* through *reactance* (*see* Chapter 3), although

this does not actually consume power. Its main influence is on frequency response of RC circuits.

Simple Coupling

Coupling one stage of a radio receiver to the next stage via a capacitor is common design practice. Although the capacity is apparently used on its own, it is associated with an effective series resistance represented by the 'load' of the stage being fed — Fig. 33. This, together with the capacitor, forms an RC combination which will have a particular time constant. It is important that this time constant matches the requirements of the *ac* signal *frequency* being passed from one stage to the other.

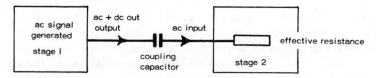

Fig. 33. Basic function of a coupling capacitor is to pass ac *signals and block* dc *signals. It will also pass undulating* dc *signals*

In the case of AM radio stage, the maximum af signal likely to be present is 10 kHz. The 'cycle' time of such a signal is $1/10000 = 0.1$ milliseconds. However, to pass this frequency each cycle represents two charge/discharge functions as far as the coupling capacitor is concerned, one positive and one negative. Thus the time period for a single charge/discharge function is 0.05 milliseconds.

The RC *time constant* necessary to accommodate this working needs to be this value to 'pass' 63 per cent of the applied *ac* voltage — and preferably rather less to 'pass' more than 63 per cent of the applied voltage.

These figures can give a clue as to the optimum value of coupling capacitor to use. For example, the typical input resistance of a low power transistor is of the order of 1000 ohms. The time constant of a matching RC coupling would be 0.05 milliseconds (*see* above), giving the requirement:

$$0.05 \times 10 = 1000 \times C$$
$$\text{or } C = 0.05 \times 10^{-9} \text{ farads}$$
$$= 0.50 \text{ pF (or preferably rather less, since this}$$
would ensure more than 63 per cent voltage 'passed').

In practice, a much higher capacitance value would normally be used, e.g. even as high as 1 μF or more. This will usually give better results, at the expense of efficiency of *ac* (in this case *rf*) transmission. (An apparent contradiction, but it happens to work out that way because the load is *reactive* rather than purely resistive). What simple calculation does really show is that capacitive coupling becomes increasingly less efficient with increasing frequency of *ac* signal when associated with practical values of capacitors used for coupling duties.

Filter Circuits

A basic RC combination used as a *filter circuit* is shown in Fig. 34. From the input side this represents a resistor in series with a capacitive *reactance*, with a voltage drop across each component. If the reactance of the capacitor (X_C) is much greater than R, most of the input voltage appears across the capacitor and thus the output voltage approaches the input voltage in value. Reactance is inversely proportional to frequency, however, and so with increasing frequency the reactance of the capacitor decreases, and so will the output voltage (an increasing proportion of the input voltage being dropped by the resistor).

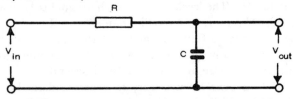

Fig. 34. Basic filter circuit. It will block ac frequencies higher than the cut-off frequency of the combination of R and C

As far as effective passage of *ac* is concerned there is a critical frequency at which the reactance component becomes so degraded in value that such a circuit starts to become blocking rather than conductive, i.e. the ratio of volts$_{out}$/volts$_{in}$ starts to fall rapidly. This is shown in simplified diagrammatic form in Fig. 35. The critical point, known as the 'roll-off' point or *cut-off frequency* (f_c) is given by

$$f_c = \frac{1}{2\pi\,RC}$$

where R is in ohms
C is in farads
$\pi = 3.1416$

But RC, as noted previously is equal to the time constant of the RC combination.

Hence:

$$f_c = \frac{1}{2\pi T}$$

where T is the time constant, in seconds.

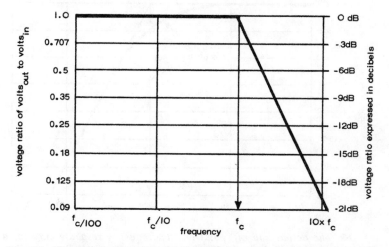

Fig. 35. Simplified diagram of how the ratio of volts in/volts out drops rapidly as the cut-off frequency of a filter is exceeded. All signals below the cut-off frequency are passed without attenuation

The performance of such a filter circuit is defined by its cut-off frequency and the rate at which the $volts_{in}$/$volts_{out}$ratio falls above the cut-off frequency. The latter is normally quoted as — (so many) dB per octave (or each doubling of frequency) — *see* Fig. 36 which shows the relationship between dB and $volts_{in}$/$volts_{out}$ ratio; and also the true form of the frequency response curve.

Circuits of this type are called *low pass filters* because they pass *ac* signals below the cut-off frequency with little or no loss or *attenuation* of signal strength. With signals above the cut-off frequency there is increasing attenuation. Suitable component values are readily calculated. For example, a typical *scratch filter* associated with a record player or amplifier used with a record deck would be designed to attenuate frequencies above, say, 10 kHz — Fig. 37. This value represents the cut-off frequency required,

$$\text{i.e. } 10\,000 = \frac{1}{2\pi\,RC}$$

$$\text{or } RC = 1600$$

Any combination of R (in ohms) and C (in farads) giving this product value could be used.

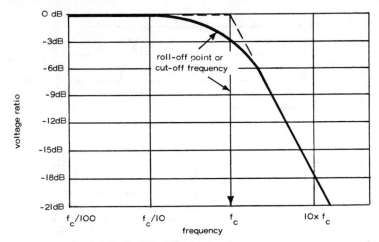

Fig. 36. The actual 'roll off' point on the frequency response curve of a filter is not sharply defined. The cut-off frequency is really a nominal figure and generally taken as the frequency at which there is a 3 decibel loss or a volts in/volts out ratio of 0.707. This is equivalent to a 50 per cent loss of power

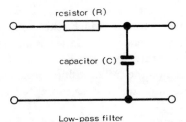

Fig. 37. Simple scratch filter circuit. Any combination of component values giving a product of RC = 1600 will work

High-Pass Filters

High-pass filters work the other way round. They attenuate frequencies below the cut-off frequency, but pass frequencies at and above the cut-off frequency with no attenuation. To achieve this mode of working the two components in the circuit are interchanged — Fig. 38.

This type of filter is again commonly associated with record player circuits, incorporated to eliminate low frequency noise or 'rumble' which may be present. The design cut-off frequency must be low

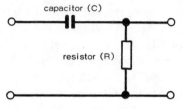

High-pass filter

Fig. 38. (right) High pass or 'rumble' filters cut off low frequencies but pass high frequencies. A typical value for the product RC in this case would be 125

enough not to interfere with bass response, and so the value chosen is usually of the order of 15 to 20 Hz. Exactly the same formula is used to determine the cut-off frequency, hence using a design value of 20 Hz:

$$20 = \frac{1}{2\pi\, RC}$$

or

$$RC = 125$$

Again any combination of R (in ohms) and C (in farads) giving a product of 125 would work.

In practical circuits such filters are normally inserted in the pre-amplifier stage, or in the amplifier immediately in front of the tone control circuit. For Hi-Fi systems the type of filter circuits used are considerably more complicated than the ones described.

Tuned Circuits

Another very important application of capacitors is in *tuned circuits*. Here a capacitor is associated with an *inductance*, a subject which is dealt with in the next chapter.

CHAPTER 7

Coils and Inductances

The flow of electric current through any conductor has the effect of generating a magnetic field. This creation of magnetic energy represents a power loss during the time that field is being created, which is measurable in terms of a 'voltage drop' or *back emf*. This is quite different (and additional) to the voltage drop produced by the *resistance* of the conductor, and disappears once stable conditions have been reached. Thus in a *dc* circuit, the *back emf* tends to prevent the current rising rapidly when the circuit is switched on. Once a constant magnetic field has been established the *back emf* disappears since no further energy is being extracted from the circuit and transferred to the magnetic field.

In the case of an *ac* circuit the current is continually changing, creating a *back emf* which is also changing at a similar rate. The value of the *back emf* is dependent both on the rate of change of current (i.e. frequency) and to a factor dependent on the form of the conductor which governs its *inductance*. Inductance is thus another form of resistance to *ac*, generated in addition to the pure resistance.

Every conductor has inductance when carrying *ac*, although in the case of straight wires this is usually negligible (except at very high frequencies). If the wire is wound in the form of a coil, however, its inductance is greatly increased. If the coil is fitted with an iron core, then its inductance will be even higher for the same number of turns and coil size.

With *ac* flowing through the coil the 'resistive' condition established is not as drastic as may appear at first sight. The polarity of the *back emf* is always such as to oppose any change in current. Thus whilst the current is increasing, work is being done against the *back emf* by storing energy in the magnetic field. On the next part of the current cycle when the current is falling, the stored energy in the magnetic field returns to the circuit, thus tending to keep the current flowing e.g. *see* Fig. 39. An inductance, in fact, may be a very good conductor of *ac*, especially when combined with a capacitor in a *tuned circuit* (see later). On the other hand it may be designed to work as a 'resistive' component or *choke*.

The inductance of a single-layer coil, wound with space between adjacent turns can be calculated from the formula:

$$L = \frac{R^2 N^2}{9R + 10L}$$

where L is the inductance in microhenrys
        R is the radius of the coil in inches
        N is the number of turns
        L is the length of the coil in inches.

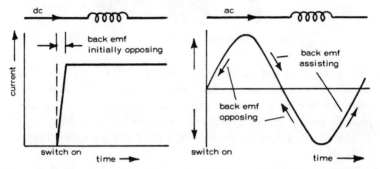

Fig. 39. Back emf *induced in a* dc *circuit on switching on exists only momentarily. In an* ac *circuit the back* emf *is continually changing*

Written as a solution for the number of turns required to produce a given inductance with R and L predetermined

$$N = \sqrt{\frac{(9R + 10L) \times L}{R^2}}$$

This formula applies regardless of the actual diameter of the wire used (also it does not matter whether bare wire or insulated wire is used), provided the *coil* diameter is very much larger than the *wire* diameter. For practical sizes of wires used for coil winding, this means a minimum coil diameter of at least 1 inch (25mm).

For smaller diameter coils the wire size will have an increasing modifying effect on the actual inductance, and even the length of leads at the ends of the coil can upset the 'formula' calculation. Thus such coils are normally designed on empirical lines (i.e. based on a specified number of turns of a given size of wire known to produce a given inductance when wound on a specific former diameter).

In practice, small coils are normally wound on a former intended to take an iron core. The position of this core is adjustable, relative to

the wound coil, by screwing in or out. Thus the actual value of
inductance can be varied, for 'tuning' purposes — Fig. 40 (*left*).

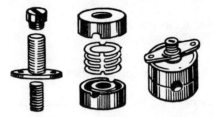

Fig. 40. Coil former (left) *and pot cores* (right)

Alternatively a *pot core* may be used where the coil is wound on a
former or bobbin, .subsequently enclosed in an iron housing. Provided the *specific inductance* of the pot core is known (it is usually
specified by the manufacturer), the number of turns (μ) to be used
for the winding can be calculated with good accuracy from the
formula:

$$N = \sqrt{\frac{L}{A_L}}$$

where L is the inductance required

        A_L is the quoted specific inductance of the pot core in
the same units as L.

Practical values of inductance used in electronic circuits may range
from microhenrys (in medium and high frequency circuits), to milli-
henrys (in low frequency circuits), up to several henrys for chokes in
power supply circuits. Normally an inductance will be wound from
the largest diameter enamelled wire it is convenient to use (and still
get the required number of turns on the former or bobbin), because
this will minimize ohmic *resistance* and thus improve the efficiency
or *Q-factor* of the coil.

Resonant Circuits
A coil (inductance) and a capacitor connected in series across an *ac*
supply has the very important characteristic that it is possible for the
reactive effect of one to cancel out the reactive effects of the other.
Thus in the 'demonstration' circuit shown in Fig. 41, L is the inductance, and C the capacitor connected across a source of *ac*, the fre-

quency of which can be varied. A resistor (R) is shown in series with L and C, as an inevitable part of a practical circuit.

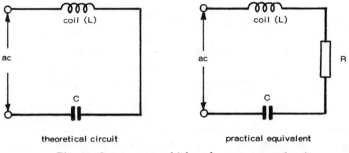

theoretical circuit   practical equivalent

Fig. 41. Components which make a resonant circuit

If the *ac* supply is adjusted to a *low* frequency, the capacitive reactance will be very much larger than R, and the inductive reactance will be much lower than R (and thus also very much lower than the capacitive reactance). See Chapter 3 for the formulas for capacitive and inductive reactance, and how these values are dependent on frequency.

If the *ac* supply is adjusted to a *high* frequency the opposite conditions will apply — inductive reactance much larger than R, and capacitive reactance lower than R and L. Somewhere between these two extremes there will be an *ac* frequency at which the reactances of the capacitance and inductance will be equal, and this is the really interesting point. When inductive reactance (X_L) equals capacitive reactance (X_C), the voltage drops across these two components will be equal but *180 degrees out of phase*. This means the two voltage drops will cancel each other out, with the result that only R is effective as total resistance to current flow. In other words, maximum current will flow through the circuit, determined only by the value of R and the applied *ac* voltage.

Working under these conditions the circuit is said to be *resonant*. Obviously resonance will occur only at a specific frequency, which is thus called the *resonant frequency*. Its value is given by the simple formula:

$$f = \frac{1}{2\pi\sqrt{LC}}$$

where f = resonant frequency in Hz
      L = inductance in henrys
      C = capacitance in farads

A more convenient formula to use is:

$$f = \frac{10^6}{2\pi\sqrt{LC}}$$

where f = resonant frequency in kilohertz (kHz)
     L = inductance in microhenrys (μH)
     C = capacitance in pico farads (pF)

Note that the formula for resonant frequency is not affected by any resistance (R) in the circuit. The presence of resistance does, however, affect the Quality factor or Q of the circuit. This is a measure of how *sharply* the circuit can be tuned to resonance, the higher the value of Q the better, in this respect. The actual value of Q is given by:

$$Q = \frac{X}{R}$$

where X is the reactance in ohms of *either* the inductance *or* capacitance at the resonant frequency (they are both the same, so it does not matter which one is taken) and R is the value of the series resistance in ohms.

The practical resonant circuit (or tuned circuit) is based on just two components — an inductance and a capacitor. Some resistance is always present, however. At low to moderately high frequencies, most of this resistance will come from the wire from which the coil is wound. At very much higher frequencies, the majority of the resistance may come from the frequency energy loss in the capacitor.

Tuned Circuits
The combination of an inductance and capacitance in series is the standard form of *tuned circuit* used in almost every radio receiver. It is drawn as shown in Fig. 42. At first, this would appear to show the coil and capacitor in parallel connection. However, the effective circuit is the 'loop', which means that the coil and capacitor are effectively in *series*.

To make the circuit tunable over a range of resonant frequencies, either component can be a variable type. The usual choice for aerial circuits is to make the capacitor variable. In practice the coil may also have variable characteristics. It is usually wound on a sleeve fitted on a ferrite rod and capable of being slid up and down the rod,

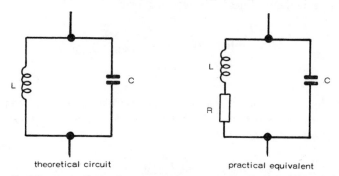

theoretical circuit   practical equivalent

Fig. 42. Theoretically, only a capacitor and inductance are involved in a resonant circuit. In practice, some resistance is always present as well

providing a means of varying the effective inductance. Once an optimum position has been found for the coil, it is cemented to the rod. In other words, the 'variable' characteristics of the coil are used only for initial adjustment. After that, all adjustment of resonant frequency or 'tuning' is done by the variable capacitor.

To assist in selecting suitable component values, the resonant frequency formula can be rewritten:

$$LC = \frac{10^{12}}{4\pi^2 \, f^2}$$

where L is in microhenrys
C is in picofarads
and f is the frequency in kHz

Maximum values of variable capacitor used are normally 300 pF or 500 pF. The working formula for calculating a matching inductance value is:

$$L \text{ (microhenrys)} = \frac{10^{12}}{4\pi^2 \, f^2 \, C}$$

As an example, suppose the tuned circuit is to be designed to cover the medium wave band, or frequencies from 500 to 1600 kHz; and a 500 pF tuning capacitor is to be used. It follows from the resonant frequency formula that maximum capacitance will correspond to the lowest resonant frequency (with a fixed inductance), which in this case is 500 kHz.

Inserting these values in the working formula

$$L \text{ (microhenrys)} = \frac{10^{12}}{4\pi^2 \times (500)^2 \times 500}$$

$$= 200$$

Now check the *resonant frequency* when the capacitor is turned to its minimum value (which will probably be about 50 pF, associated with this value of inductance:

$$f = \sqrt{\frac{10^{12}}{4\pi^2 \, LC}}$$

$$= \sqrt{\frac{10^{12}}{4\pi^2 \times 200 \times 50}} = 1600\text{kHz}$$

This shows that a 50–500 pF variable capacitor will 'tune' the circuit from 1500 kHz (the highest frequency), down to 500 kHz satisfactorily. In other words, it covers the whole of the medium wave broadcast band.

If the final results achieved in the circuit do not provide quite the coverage required, for example a station near one end of the band is not picked up, then there is still the possibility of 'shifting' the frequency coverage in one direction or the other by adjusting the inductance (i.e. sliding the coil up or down the ferrite rod).

There are other types of tuned circuits which normally require adjustment only when initially setting up. These normally employ a 'tunable' inductance (e.g. a coil wound on a former with an adjustable iron dust core). Such circuits may also be tuned by a 'trimmer' capacitor; or both a 'trimmer' capacitor and 'tunable' inductance. The latter combination provides 'double tuning'.

Parallel Resonant Circuits
Fig. 43 shows a *parallel resonant circuit*. It differs from the series resonant circuit in having the resistance effectively *paralleled* across C and R. Again it has a resonant frequency when the reactances of C and L are equal, but at this frequency the current in the line circuit is a *minimum*.

An important application of this type of resonant circuit is for *impedance matching*.

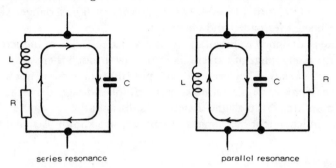

series resonance   parallel resonance

Fig. 43. Series and parallel resonance circuits. Although these may look the same, it is the 'loop' formed by L and C which determines the type of circuit and whether the resistance component (R) is in series or parallel with the 'loop'

Radio Frequency Chokes

A radio frequency choke (*rfc*) is a coil or inductance so designed that it has a relatively low ohmic resistance but a very high reactance at radio frequencies. It can thus pass *dc* but will block high frequency *ac* when the two are present in the same circuit — Fig. 44. In other words it really works the opposite way round to a capacitor as a circuit element in this respect.

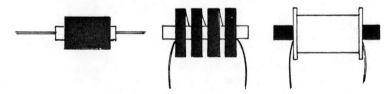

Fig. 44. Typical appearance of chokes wound on a ferrite core

The characteristics of any *rfc* will vary with frequency. At high frequencies it will have characteristics similar to that of a parallel-resonant circuit; and at low frequencies characteristics similar to that of a series-resonant circuit. At intermediate frequencies it will have intermediate characteristics. The actual characteristics are relatively unimportant when an *rfc* is used for *series feed* because the *rf* voltage across the choke is negligible. If used for *parallel feed* (where the choke is shunted across a tank circuit), it must have sufficiently high

impedance at the lowest frequencies and no series-resonance characteristics at the higher frequencies in order to reduce power
absorption to a suitable level. Otherwise there will be a danger of the
choke being overloaded and burnt out.

Chokes designed to maintain at least a critical value of inductance
over the likely range of current likely to flow through them are called
swinging chokes. They are used as input filters on power supplies to
reduce 'ripple' or residual *ac* content. Chokes designed specifically
for smoothing 'ripple' and having a substantially constant inductance, independent of changes in current, are known as *smoothing
chokes*.

CHAPTER 8

Transformers

A transformer consists of two coils so positioned that they have mutual inductance. This magnetic 'coupling' effect can be further enhanced by winding the two coils on a common iron core. The coil which is connected to the source of supply is called the *primary* (winding); and the other coil is called the *secondary* (winding). In order to 'work' or transfer electrical energy from primary to secondary, the magnetic field must be continually changing, i.e. the supply must be *ac*.

One of the most useful characteristics of a transformer is its ability to provide step-down (or step-up) of *ac* voltages. The step-down (or step-up) ratio will be proportional to the number of turns in each coil, i.e.

$$V_s = n_s/n_p \times V_p$$

where V_s = secondary voltage
n_s = number of turns on secondary
n_p = number of turns on primary
V_p = primary voltage.

The currents flowing in the primary and secondary follow a similar relationship, but in opposite ratio

$$I_s = n_p/n_s \times I_p$$

where I_s = secondary current
I_p = primary current.

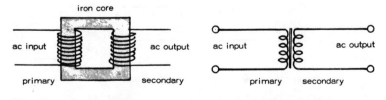

diagrammatic form   symbolic form

Fig. 45. The simple iron cored transformer

In other words, a step-down in voltage produces a step-up in current, and vice versa.

In practice there will always be some losses due to the resistance of the coils and energy lost in hysteresis and eddy currents in the core (in the case of an iron-cored transformer); and also from *reactance* caused by a leak of inductance from both coils. Thus the power which can be taken from the secondary is always less than the power put into the primary, the ratio of the two powers being a measure of the *efficiency* of the transformer.

Typically, efficiency may range from 60 per cent upwards, but is not necessarily constant. A transformer is usually designed to have its maximum efficiency at its *rated power output*. Its actual efficiency figure will decrease if the output is higher or lower. This loss of power appears in the form of heat. Thus overloading a transformer can both reduce its efficiency and increase the heating effect. Operating at reduced output has no harmful effect, except for reducing efficiency as the actual power loss (and thus heating effect) is lowered.

Transformers as Power Supplies
By selecting a suitable turns ratio a transformer can be used directly to convert an *ac* supply voltage into a lower (or higher) *ac* output voltage at efficiencies which may be as high as 90 per cent Fig. 46. There are also applications where a 1:1 turns ratio transformer is used, providing the same *ac* output voltage as the *ac* input voltage, e.g. where it is desirable to 'isolate' the supply from the output circuit. All transformers do, of course, provide physical separation of input and output circuits, but the degree of isolation safety is very much dependent on the actual construction of the transformer.

The more usual power supply application of a transformer is to step-down an *ac* voltage into some lower *dc* voltage output. The

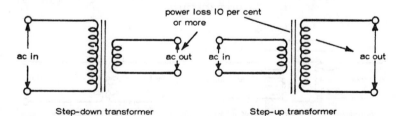

Fig. 46. Step-down and step-up transformers defined. In practice, transformers are often drawn in symbolic form with both coils of the same length, regardless of actual turns ratio

transformer will only provide voltage conversion. Additional components are needed in the output circuit to transform the converted *ac* voltage into a *dc* voltage.

Two basic circuits for doing this are shown in Fig. 47. The first uses a single diode and provides *half wave rectification*, passing one half of each *ac* cycle as *dc* and suppressing the other half cycle. The purpose of the capacitor is to maintain the *dc* voltage output as far as possible by discharging on each 'suppressed' half cycle, and for this a large value capacitor is required. Although a very simple circuit, it has the inherent disadvantage of generating high *peak* voltages and currents, especially if a high current is drawn from the output. Also the *dc* output is far from smooth. It will have a 'ripple' at the *ac* frequency.

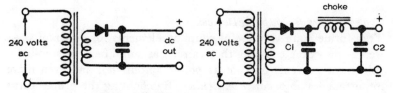

Fig. 47. Half wave rectification of ac

Much can be done to smooth the output by adding an inductance or choke and a second capacitor, as shown in the second diagram. These two components work as a filter (*see also* Chapter 6). The design of the choke has to be specially matched to the requirements, offering low resistance to *dc* without becoming 'saturated', which could reduce its inductance. In particular circuits the inductance may be a *swinging choke*, when it is possible to eliminate the reservoir capacitor CI.

The first diagram of Fig 48 shows a simple *full wave rectifier* circuit added to the transformer (the secondary of which must be centre tapped). For the same secondary voltage as the half-wave rectifier, the *dc output* voltage is now halved, but the current which can be drawn for a given rectifier rating is doubled. The reservoir capacitor charges and discharges alternately. This will produce a smoother *dc* supply, but 'ripple' will still be present and in this case is equal to *twice* the *ac* frequency.

The more usual form of full-wave rectifier is the *bridge circuit*, shown in the second diagram. This gives approximately the same no-load voltage as a half-wave rectifier with the advantage of full-wave rectification and better smoothing.

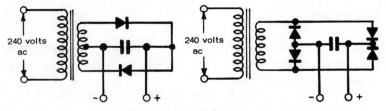

Fig. 48. Full wave rectification of ac

A practical circuit of this type is shown in Fig. 49. A single high value electrolytic capacitor is used for smoothing. Additional smoothing between stages fed from such a power supply may be provided by a resistor, associated with a decoupling capacitor (like Fig. 47). The resistor value can also be chosen to 'drop' a specific amount of voltage if the previous stage(s) do not require the full power supply output voltage.

Transformers as Coupling Devices

Transformers are very useful coupling elements for *ac* circuits. As well as providing coupling they can act as 'amplifiers' to 'step-up' a voltage or current (but not as *power* amplifiers); and even more important for *impedance matching*. By choosing the proper turns ratio the impedance of a fixed load can be transformed to any desired higher or lower impedance, within practical limits. This can be a particularly important requirement when coupling transistor radio stages.

For impedance matching, the following relationship applies:

$$\frac{N_p}{N_s} = \sqrt{\frac{Z_P}{Z_S}}$$

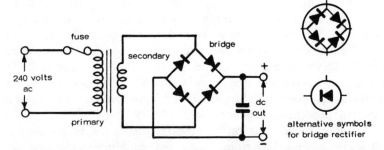

Fig. 49. Practical power supply circuit. A high value capacitor is used. The four diodes are bought as a single component called a bridge rectifier

where Z_p is the impedance of the transformer looking into the primary terminals

where Z_s is the impedance of the load connected to the secondary of the transformer.

For impedance matching it is therefore necessary to design the primary to provide the required Z_p and select the turns ratio to satisfy the equation.

Autotransformers

An *autotransformer* is a one-winding coil with an intermediate tapping point. The full length of the coil (usually) forms the primary, and the length of coil between the tapping point and one end of the coil the secondary — Fig. 50. It works on exactly the same principle as a conventional transformer, with the voltage developed across the output proportional to the turns ratio of this length of coil to the full length of coil. This common length of coil effectively has the primary current and load current flowing through it in *opposite* directions, so that the resultant current is the difference between them.

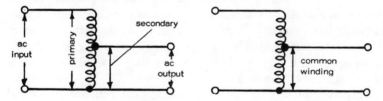

Fig. 50. The autotransformer is a single full-length coil with a tapping point

CHAPTER 9

Semiconductors: Diodes, Transistors and FETs

Resistors, capacitors and inductances are known as *passive* components. Devices which produce changes in circuit conditions by reacting to applied signals are known as *active* components. The majority of active components used in modern electronic circuits are *semiconductors*, or more correctly put, devices based on semiconductor materials.

Put very simply, a semiconductor material is one which can be given a predominance of mobile negative charges, or electrons; or positive charges or 'holes'. Current can flow through the material from the movement of both electrons and 'holes'. This is quite different from the behaviour of a normal conductor, where current flow is the result of electrons *through* the material (*see* Chapter 1).

Semiconductor properties can be given to a strictly limited number of materials by 'doping' with minute traces of impurities. The two main semiconductor materials are germanium and silicon (both non-metals or 'semi metals'). 'Doping' can produce a material with either a predominance of *P*ositive charges ('holes') resulting in a *P-type* material: or with a predominance of *N*egative charges ('electrons'), known as an *N-type* material.

This does not become particularly significant until a *single* crystal (or germanium or silicon) is treated with both a P-type dope and an N-type dope. In this case, two separate regions are formed — a P region and an N region. Since these regions have opposite charges there will be a tendency for electrons to migrate from the N-zone to the P-zone, and 'holes' to migrate from the P-zone into the N-zone. The effect will be a cancellation of charges in the region of the junction of the P and N zones, forming what is called a *depletion layer* — Fig. 51. This layer, which contains no free electrons or

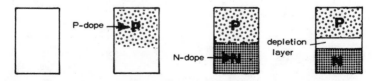

Fig. 51. Four stages in the construction of a semiconductor diode, shown in simple diagrammatic form

'holes', will then act as a barrier between the P-zone and the N-zone, preventing any further migration of either electrons or 'holes'. In effect, the barrier or depletion layer sets up a potential difference between the two regions and the device remains in a stable state until an external voltage is applied to it.

Fig. 52 shows what happens when an external voltage *is* applied to the device. In the first diagram the voltage is connected + to the P zone. Provided this voltage is sufficiently high to overcome the potential difference set up in the construction of the device (which may be only a few tenths of a volt) it will repel 'holes' in the P zone towards the N zone; and attract electrons in the N zone into the P zone. Effectively the barrier or depletion layer will disappear and current will flow through the device. Voltage applied this way round is known as *forward bias*.

If the external voltage is applied the other way round, as in the second diagram, the opposite effect is created, i.e. the thickness of the depletion layer will increase, thus building up a higher potential in the device *opposing* the external voltage. The 'back voltage' developed will be equal to that of the applied voltage, so no current will flow through the device. Voltage applied this way round is known as *negative bias*.

The device just described is a semiconductor *diode*. It has the basic characteristic of acting as a conductor when connected to an external voltage one way round (forward bias); and as an insulator when connected the other way round (reverse bias).

Diode characteristics will be described in some detail later on, but having established the electronic 'picture' of a diode, the same principles can be applied to explain the working of a *transistor*.

Basically a transistor is two diodes placed back-to-back with a common middle layer, the middle layer in this case being much

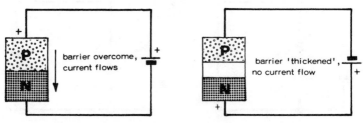

Fig. 52. The two modes in which a diode can be worked

thinner than the other two. Two configurations are obviously possible, P-N-P or N-P-N (Fig. 53). These descriptions are used to describe the two basic types of transistors. Because a transistor contains elements with two different polarities (i.e. 'P' and 'N' zones), it is referred to as a bipolar device, or *bipolar transistor*.

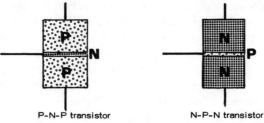

P-N-P transistor   N-P-N transistor

Fig. 53. Construction of a P-N-P and a N-P-N transistor, shown in simple diagrammatic form

A transistor has three elements, and to operate in a working circuit it is connected with two external voltages or polarities. One external voltage is working effectively *as* a diode. A transistor will, in fact, work as a diode by using just this connection and forgetting about the top half. An example is the substitution of a transistor for a diode as the detector in a simple radio. It will work just as well as a diode as it *is* working as a diode in this case.

The diode circuit can be given forward or reverse bias. Connected with forward bias, as in the first diagram of Fig. 54, drawn for a P-N-P transistor, current will flow from P to the bottom N. If a second voltage is applied to the top and bottom sections of the transistor, with the *same* polarity applied to the bottom, the electrons already flowing through the bottom N section will promote a flow of current through the transistor bottom-to-top.

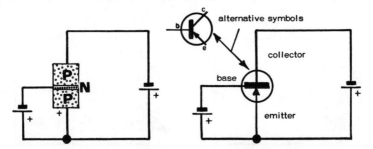

Fig. 54. Bias and supply connections to a P-N-P transistor shown diagrammatically (left) and in symbol form (right)

By controlling the degree of doping in the different layers of the transistor during manufacture, this ability to conduct current through the second circuit through the resistor can be very marked. Effectively, when the bottom half is forward biased, the bottom section acts as a generous source of free electrons (and because it emits electrons it is called the *emitter*). These are collected readily by the top half, which is consequently called the *collector*, but the actual amount of current which flows through this particular circuit is controlled by the bias applied at the centre layer, which is called the *base*.

Effectively, therefore, there are two separate 'working' circuits when a transistor is working with correctly connected polarities. One is the loop formed by the bias voltage supply encompassing the emitter and base. This is called the *base* circuit or *input* circuit. The second is the circuit formed by the collector voltage supply and all three elements of the transistor. This is called the *collector* circuit or *output* circuit. (Note: this description applies only when the emitter connection is common to both circuits — known as *common emitter* configuration. This is the most widely used way of connecting transistors, but there are, of course, two other alternative configurations — *common base* and *common emitter*. But the same principles apply in the working of the transistor in each case).

The particular advantage offered by this circuit is that a relatively small *base* current can control and instigate a very much larger *collector* current (or, more correctly, a small input *power* is capable of producing a much larger *output* power). In other words, the transistor works as an *amplifier*.

With this mode of working the base-emitter circuit is the input side; and the emitter through base to collector circuit the output side. Although these have a common path through base and

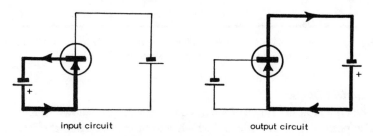

input circuit   output circuit

Fig. 55. The two separate circuits involved in working a transistor. Direction of current flow is for a P-N-P transistor

emitter, the two circuits are effectively separated by the fact that as far as polarity of the base circuit is concerned, the base and upper half of the transistor are connected as a *reverse biased* diode. Hence there is no current flow from the base circuit into the collector circuit.

For the circuit to work, of course, polarities of both the base and collector circuits have to be correct (forward bias applied to the base circuit, and the collector supply connected so that the polarity of the common element (the emitter) is the same from both voltage sources). This also means that the polarity of the voltages must be correct for the type of transistor. In the case of a P-N-P transistor, as described, the emitter voltage must be *positive*. It follows that both the base and collector are negatively connected with respect to the emitter. The symbol for a P-N-P transistor has an arrow on the emitter indicating the direction of *current flow*, i.e. always *towards* the base. ('P' for positive, with a *P*-N-P transistor).

In the case of an N-P-N transistor, exactly the same working principles apply but the *polarities* of both supplies are reversed — Fig. 56. That is to say, the emitter is always made negative relative to base and collector ('N' for 'negative' in the case of an N-P-N transistor). This is also inferred by the reverse direction of the arrow on the emitter in the symbol for an N-P-N transistor, i.e. current flow *away* from the base.

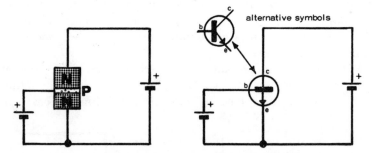

Fig. 56. Bias and supply connections to an N-P-N transistor, shown diagrammatically (left) and in symbolic form (right)

Practical Diodes

The typical appearance of a semiconductor diode is shown in Fig. 57. The positive end (i.e. the end which is to be connected to the positive side of a circuit) is usually marked by a red dot or colour band, or a + sign; and also usually with a type number consisting of one or

more letters followed by figures. This identifies the diode by manufacturer and specific 'model'. Specific type numbers are usually
quoted for specific circuit designs, but many circuits are fairly non-
critical as regards the type of diode used.

Diodes may also be described in more general terms by the crystal
material (germanium or silicon); and by construction. Here choice
can be more important. Germanium diodes start conducting at lower
voltages than silicon diodes (about 0.2 to 0.3 volts, as compared
with 0.6 volts); but tend to have higher leakage currents when reverse
biased, this leakage current increasing fairly substantially with
increasing temperature. Thus the germanium diode is inherently less
efficient as a rectifier than a silicon diode, especially if reverse bias
current is high enough to produce appreciable heating effect. On the
other hand a germanium diode is preferred to a silicon diode where
very low 'operating' voltages are involved because it starts to conduct
at a lower forward voltage.

The construction of a diode governs both its current-carrying
capabilities when conducting, and its capacitance effect. The larger
the *junction area* of a diode, the higher the current it can pass
without overheating — this characteristic being desirable in high-
power rectifiers, for example. On the other hand, increasing the
junction area increases the readiness with which a diode will pass *ac*
due to inherent capacitance effects. To reduce this effect to a minimum, a diode can be made from a single 'doped' crystal (usually N-
type), on which the point of a piece of spring wire rests. The end of
this wire is given opposite doping (i.e. P-type). This reduced the

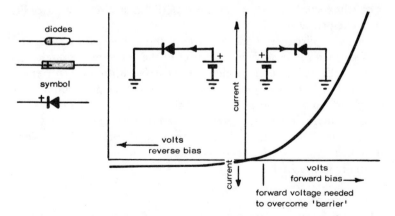

Fig. 57. Working characteristics of a typical semiconductor diode

junction area to a minimum, such as a diode being known as a *point-contact* type. It is a favoured type for use in circuits carrying high frequency *ac* signals, and for this reason is sometimes called a *signal diode*.

The typical characteristics of a diode are also shown in simple graphical form in Fig. 57. Bias is represented by the voltage applied to the positive side, specifically referred to as *anode voltage*. Current flowing through the diode is referred to as *anode current*.

With *forward bias* (positive voltage applied to the + end of the diode), there will be at first no anode current until the inherent barrier voltage has been overcome (e.g. 0.3 volts for a germanium diode; 0.6 volts for a silicon diode — regardless of the *construction* of either type). Any further increase in anode voltage will then produce a steep rise of anode current. In practice it is necessary to limit this current with a resistor or equivalent resistive load in the circuit to prevent the diode being overheated and the junction destroyed.

With reverse bias (negative voltage applied to the + end of the diode), the only current flowing will be a very small leakage current of the order of microamps only and normally quite negligible. This leakage current does not increase appreciably with rise in (negative) anode voltage, once it has reached its saturation value.

It will be appreciated that a diode will work in both a *dc* and an *ac* circuit. In a *dc* circuit, it will conduct current if connected with forward bias. If connected the opposite way round, it will act as a 'stop' for current flow. An example of this type of use is where a diode is included in a *dc* circuit — say the output side of a *dc* power supply — to eliminate any possibility of reverse polarity voltage surges occurring which could damage transistors in the same circuit (e.g. *see* Fig. 133, Chapter 20).

In an *ac* circuit a diode will 'chop' the applied *ac*, passing half cycles which are positive with respect to the + end of the diode, and stopping those half cycles which are negative with respect to + end of the diode. This is *rectifier* action, widely used in transforming an *ac* supply into a *dc* output. The same action is required of a *detector* in a radio circuit. Here the current applied to the diode is a mixture of *dc* and *ac*. The *diode detector* transforms this mixed input signal into a *varying dc* output, the variations following the form of the *ac* content of the signal.

Basic Transistor Circuits
The transistor in common-emitter configuration works as an

amplifier, as previously explained. It needs two separate supply voltages — one for bias and the other for the collector — but these do not necessarily have to come from separate batteries. They can be provided by a single supply (battery) taken to the common connection (the emitter) and the collector; and tapping the collector side to apply the necessary forward bias voltage to the base dropped through a bias resistor.

A basic amplifier circuit then looks like Fig. 58. To make the circuit do useful work, the collector current has to be fed through an output load, e.g. a load resistor. These two diagrams also show clearly input and output as separate entities, and can clarify the point about amplification. The power derived in the output is far greater than that put into the input.

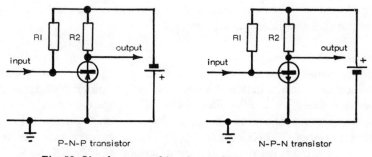

Fig. 58. Simple current bias circuits for transistor working

This very simple method of supplying both collector and bias voltages from a single source is known as *current biasing*. It needs only one resistor, and it works. The resistor value is chosen to give a base-emitter voltage of the order of 0.1 to 0.2 volts for germanium transistors; and about 0.6 to 0.7 for silicon transistors. It is not as stable as it should be for many circuits, however, particularly if a germanium transistor is used. Thus voltage bias is often preferred — see Fig. 59.

With voltage bias, two resistors (R1 and R1) are used to work as a potential divider. A resistor (R3) is also added in the emitter line to provide *emitter feedback* automatically to control the bias voltage under varying working conditions. This latter resistor is also usually paralleled with a capacitor to provide further stabilization (but this may be omitted with silicon transistors).

Determination of suitable component values is now more

complicated since three resistors are involved. The actual base voltage can be calculated from the following formula:

$$\text{base voltage} = \frac{R2}{(R1 + R2)} \times \text{supply voltage}$$

when the emitter voltage will be equal to this *less* the voltage between base and emitter (across the transistor). In most cases a voltage drop of 1 volt in the case of germanium transistors and 3 volts with silicon transistors is the design aim. The emitter resistor (R3) also needs to be quite large so that there will be minimal changes in emitter current with any variation in the supply voltage. This can cause a little re-thinking about suitable values for R1 and R2, for the voltage developed across R3 must be very much greater than the voltage developed by the base current across the *source resistance* formed by the parallel combination of R1 and R2.

Transistor Construction

The original transistors were made from germanium crystals with point-contact construction. Later types, with considerably improved performance, are of *alloy-junction* or *alloy-diffusion* construction. Silicon transistors are usually made by the·*planar* process (silicon planar process). Their characteristics can be further improved by adopting a modified planar process described as *epitaxial*, basically involving a preliminary process of forming an oriented layer (epitaxial layer) of lightly doped silicon over the silicon substrate. The transistor elements are subsequently formed within the layer rather than within the silicon substrate itself (as in the normal planar process). Epitaxial silicon planar transistors have superior charac-

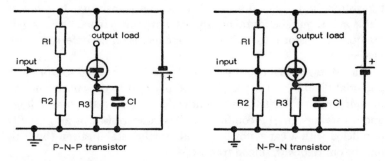

Fig. 59. Voltage bias circuits for transistor working

teristics for high frequency applications, notable in *rf* and *if* circuits for superhet radios.

Germanium and Silicon Transistors

Just like diodes, transistors are made from either germanium or silicon crystals. Germanium transistors have low voltage losses but their characteristics are more liable to vary with temperature, so that the *spread* of characteristics under which they work in a circuit can be quite wide. They are also limited to a maximum working temperature of about 100°C.

Silicon transistors are generally more stable and can operate at temperatures up to 150°C or more. They have lower leakage losses and higher voltage ratings, and are generally far better suited for use in high frequency circuits.

The Shape of Transistors

Transistors come in all sorts of shapes and sizes. However the only problem where a specified type of transistor is to be used is correctly identifying the three leads, when the position of these can be identified by reference to Fig 60. The most common lead configuration is in line, with a circular case, when the leads follow in logical order — collector, base, emitter, with the collector lead being more widely spaced from the middle (base) lead than the emitter lead, looking at the bottom of the transistor from where the leads emerge. This does

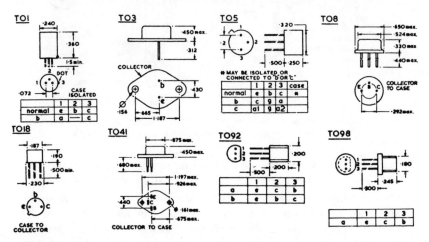

Fig. 60. Some common transistor outlines (diagrams by Electrovalue, all dimensions in inches)

not apply when the case is partly circular with a flat on one side. Here the three leads are equi-spaced and with the flat side to the left (and looking at the bottom), the lead arrangement may be bce, cbe or ebc.

Power transistors are more readily identified by their elongated bottom with two mounting holes. In this case there will only be two leads — the emitter and base — and these will normally be marked. The collector is connected internally to the can, and so connection to the collector is via one of the mounting bolts or bottom of the can.

Field Effect Transistors (FETs)

The *Field Effect transistor* (or FET) is really a different type of semiconductor device to a bipolar transistor, with characteristics more like a thermionic valve than a bipolar transistor. Its correct definition is a *unipolar* transistor. The way in which it works can be understood by presenting it in electronic 'picture' form as in Fig. 61 where it can be seen that it consists of a *channel* of either P-type or N-type semiconductor material with a collar or *gate* of opposite type material at the centre. This forms a semiconductor junction at this point. One end of the channel is called the *source*, and the other end the drain.

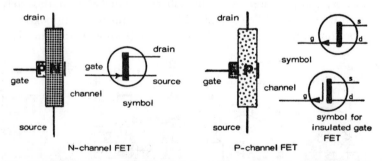

Fig. 61. Construction of field effect transistors shown in simple diagrammatic form, together with appropriate symbols for FETs

An FET is connected in a similar manner to a bipolar transistor, with a *bias* voltage applied between gate and source, and a supply voltage applied across the centre of the channel (i.e. between source and drain). The source is thus the common connection between the two circuits. Compared with a bipolar transistor, however, the bias voltage is *reversed*. That is, the N gate material of a P-channel FET is biased with *positive* voltage; and the P gate material of an N-channel FET is biased with *negative* voltage — Fig. 62. This puts the two system voltages 'in opposition' at the source, which is responsible for the characteristically *high input resistance* of FETs.

The effect of this reverse bias is to form an enlarged depletion layer in the middle of the channel, producing a 'pinching' effect on the flow of electrons through the channel and consequently on the current flow in the source-to-drain circuit. If enough bias voltage is applied the depletion layer fills the whole gate ('shuts the gate'), causing *pinch off*, when the source-to-drain current falls to zero (in practice nearly to zero, for there will still be some leakage). With *no* bias applied to the gate, the gate is 'wide open' and so maximum current flows.

In effect, then, the amount of reverse bias applied to the *gate* governs how much of the gate is effectively open for current flow. A relatively small change in gate voltage can produce a large change in source-to-drain current, and so the device works as an amplifier. In this respect a P-channel FET works very much like a P-N-P transistor, and an N-channel FET as an N-P-N transistor. Its main advantage is that it can be made just as compact in size, but can carry much more power. In this respect — and the fact that it has a high input resistance, whereas a bipolar transistor has a low input resistance — it is more like a thermionic value in characteristics than a bipolar transistor. It also has other advantages over a bipolar transistor, notably a much lower inherent 'noise', making it a more favourable choice for an amplifier in a high quality radio circuit.

The type of field effect transistor described is correctly called a junction field effect transistor, or JFET. There are other types produced by modifying the construction. The insulated gate field effect transistor or IGFET is self-explanatory. The IGFET has even higher input resistance (because the gate is insulated from the channel), and is also more flexible in application since either 'reverse'

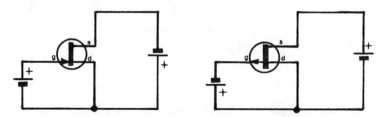

P-channel FET   N-channel FET

Fig. 62. Basic bias requirements for field effect transistors

or 'forward' polarity can be applied to the gate for bias. FETs, of either type, can also be made with two gates. In this case the first gate becomes the *signal gate* (to which the input signal is applied) and the second gate becomes the *control gate*, with similar working to a pentode valve (*see* Chapter 12).

FETs are also classified by the mode in which they work. A JFET works in the *depletion mode*, i.e. control of the extent of the depletion layer, and thus the 'gate opening' being by the application of a bias voltage to the gate. An IGFET can work in this mode, or with opposite bias polarity, in which case the effect is to produce an increasing 'gate opening', with enhanced (increased) source-to-drain current. This is called the *enhancement mode*.

An FET designed specifically to work in the enhancement mode has no channel to start with, only a gate. Application of a gate voltage causes a channel to be formed.

The basic circuit of an FET amplifier is very simple — e.g. Fig. 63 (with polarity drawn for a P-channel FET). Instead of applying a definite negative bias to the gate, a high value resistor R1 is used to maintain the gate at substantially zero voltage. The value of resistor R2 is then selected to adjust the potential of the source to the required amount *positive* to the gate. The effect is then the same as if negative bias were applied direct to the gate. This arrangement will also be self-compensating with variations in source-to-drain current. The third resistor R3 is a load resistor for the FET to set the design operating current. Capacitor C1 acts as a conductive path to remove signal currents from the source.

Both junction-type (JFET) and insulated gate (IGFET) field effect transistors are widely used, the latter having the wider application, particularly in integrated circuits. The metal-oxide semiconductor FET, generally referred to as a MOSFET, can be designed to work in

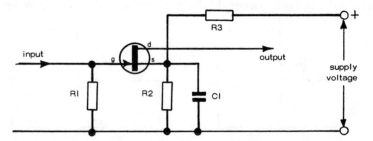

Fig. 63. Basic FET amplifier circuit. Performance is generally superior to that of a bipolar transistor amplifier

either mode, i.e. as a depletion MOSFET, or enhancement MOSFET. The formeı is usually an N-channel device and the latter a P-channel device. P-channel MOSFETS working in the enhancement mode are by far the more popular, mainly because they are easy to produce. In fact, an N-channel MOSFET can be made smaller for the same duty and has faster switching capabilities, and so really is to be preferred for LSI MOS systems (*see also* Chapter 13).

CHAPTER 10

Neons, LEDs and Liquid Crystals

The *neon* is a glow lamp consisting of an evacuated glass envelope fitted with two separated electrodes and filled with an inert gas (neon or argon). If connected to a low voltage the resistance is so high that the neon provides virtually an open circuit, but if the voltage is increased there comes a point where the gas ionises and becomes highly conductive, as well as giving off a glowing light located on the negative electrode. If the gas is neon, the glow is orange in colour. More rarely argon is used as the gas, in which case the glow is blue.

The characteristic performance of a neon is shown in Fig. 64. The voltage at which the neon starts to glow is called the *initial break-down voltage*. Once this has been reached and the bulb triggered into 'firing' (glowing), the voltage drop across the neon will remain virtually constant regardless of any increase in current in the circuit. At the same time the area of glow will increase with increasing current, up to the point where the entire surface of the negative electrode is covered by glow. Any further increase in current will then push the neon into an *arc* condition, where the glow changes to a blue-white point of light on the negative electrode and results in rapid destruction of the lamp.

To operate a neon successfully, therefore, it is necessary to have enough voltage for the neon to 'fire', and after that enough resistance in the circuit to limit the current to that which will ensure that the

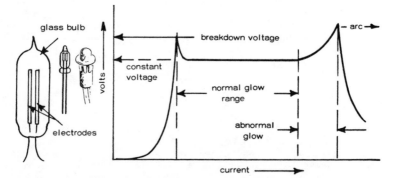

Fig. 64. Typical neon construction and characteristic performance. A neon only works in a dc circuit

neon remains operating in the normal glow region. Because the resistance of the neon itself is very low after 'firing', this means the use of a resistor in series with the neon, known as a *ballast resistor*. Typically the 'firing' or breakdown voltage may be anything from about 60 volts to 100 volts (or in some cases even higher). The continuous current rating is quite low, usually between 0.1 and 10 milliamps. The series resistor value is chosen accordingly, related to the voltage of the supply to which the neon will be connected. In the case of neons to be operated off a 250 volt (mains) supply, a 220 k ohm resistor is normally adequate — Fig. 65. With some commercial neons the resistor may actually be built into the body of the neon.

Lacking any specific information on this subject it can be assumed that a neon will have no resistance when glowing, but will drop 80 volts. A suitable value for a ballast resistor can be calculated on this basis, related to the actual voltage of the supply to be used, and assuming a 'safe' current of, say 0.2 milliamps, for example:

For 250 volt supply, resistor has to drop $250 - 80 = 170$ volts.
Current through resistor and neon (in series) is to be 0.2 mA

$$\text{Therefore resistance} = \frac{\text{volts}}{\text{amps}}$$

$$= \frac{170}{0.2 \times 1/1000}$$

$$= 850 \text{ k ohms, or say 1 megohm.}$$

This should be playing safe with most commercial neons. If the glow is not very bright, the value of the ballast resistor can be decreased to operate the neon farther along the normal glow region. But the resistance should never be decreased so much that the whole of the negative electrode is covered by glow as this will indicate that the neon is becoming overloaded and approaching the arc condition.

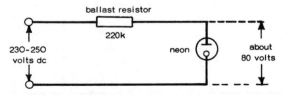

Fig. 65. In a practical circuit a neon is always connected in series with a ballast resistor to limit current flow

Another point about the strength of the glow light is that it will normally appear *brighter* in light than in dark. In fact, in complete darkness the glow may be erratic and/or require a higher breakdown voltage to start it. Some neons have a minute trace of radioactive gas added to the inert gas to stimulate ionization, when this particular effect will not be noticeable.

Because of the 'constant voltage' characteristics of a neon under normal glow conditions, it can be used as a voltage stabilizing device. Thus in the circuit shown in Fig. 65 the output tapped from each side of the neon will be a source of constant voltage as long as the neon remains working in the normal glow region. This voltage will be the same as the nominal breakdown voltage of the neon.

The use of a neon as a *flasher* in a relaxation oscillator circuit has already been described (Fig. 31, Chapter 6). A variation on this is shown in Fig. 66, using a 1 megohm potentiometer as the ballast resistor and two 45-volt or four 22½-volt dry batteries as the source of supply. The potentiometer is adjusted until the neon lights up. The control is then turned the other way until the neon just goes out. Leaving the potentiometer in this position, the neon should then flash at regular intervals determined by the value of the capacitor.

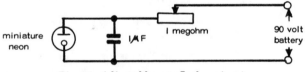

Fig. 66. Adjustable rate flasher circuit

An adaption of this circuit is shown in Fig. 67, where the circuit is switched by a Morse key. Phones can be connected across the point shown to listen into the Morse signals, which are also visible as a flashing light. An ordinary bulb would work just as well as a visible indicator (and with a much lower voltage required), but in this case

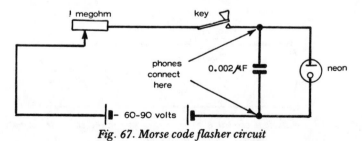

Fig. 67. Morse code flasher circuit

the signals would only be heard as 'clicks'. With the neon circuit, the actual oscillation of the relaxation oscillator is heard. The *time constant* of the circuit is governed by the value of the capacitor and the setting of the ballast potentiometer.

A further extension of the use of a neon as an *oscillator* in a relaxation oscillator circuit is shown in Fig. 68. This is a true signal generator circuit, the output of which should be audible in headphones or even a small loudspeaker with the 'tone' adjustable by the potentiometer.

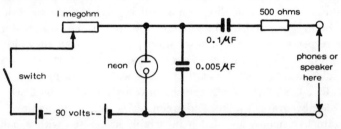

Fig. 68. Simple tone generator based on an NE-2 miniature neon

Neon 'flashers' can be made to work in random fashion (again *see* Chapter 6), or sequentially. A circuit for a sequential flasher is shown in Fig. 69. More stages can be added to this circuit, if desired, taking the connection of C3 to the last stage.

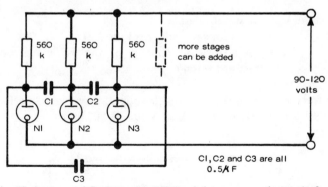

Fig. 69. Sequential flasher using NE-2 miniature neons (or equivalent)

Finally an *astable multivibrator* circuit is shown in Fig. 70, using two neons. These will flash on and off in sequence at a rate determined by R1 and R2 (which should be equal in value) and C1.

As a general guide to 'flasher' timing, *increasing* the value of the ballast resistor or the capacitor in the relaxation oscillator circuit will *slow* the rate of flashing; and vice versa. To preserve the life of a typical neon, however, the value of ballast resistor used should not be less than about 100 k ohms; and best results in simple relaxation oscillator circuits will usually be achieved by keeping the capacitor value below 1 microfarad.

LEDs

LED is short for Light Emitting Diode. This is essentially a two-element semi-conductor device where the energy produced by conduction in a specific direction is radiated as light, the intensity of light being governed by the current flowing through the diode. In these respects they are somewhat similar to neons, but they light at very much lower forward voltages (typically 1.6 to 2 volts) and can generally draw higher forward currents without burning out (e.g. typically 20 mA). Originally the colour of light emitted by LEDs was red, but now orange, yellow and green LEDs are also available.

Again like neons, an LED is invariably associated with a ballast resistor in series to limit the voltage applied to the LED and the current flowing through it. The value of resistor required is:

$$R = \frac{V_s - V_f}{I_f}$$

where V_s = *dc* supply voltage
V_f = rated forward voltage of the LED
I_f = rated forward current of the LED at specified forward voltage.

Thus for operating off, say, a 6 volt supply, a typical value for the ballast resistor would be $(6 - 2)/20 \times 10^{-3} = 200$ ohms.

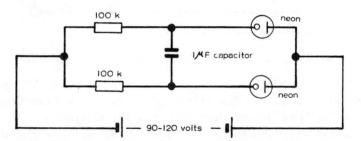

Fig. 70. Astable multivibrator circuit, each neon flashing in turn

In the case of an *ac* supply, a diode is connected in inverse parallel with the LED and the resistor value required is *one half* that given by the above formula — *see also* Fig. 71.

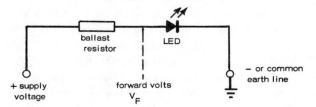

Fig. 71. LEDs are invariable connected with a ballast resistor in series to drop the supply voltage to the required forward voltage. Note the symbol for an LED (light emitting diode)

LEDs are very familiar in the form of groups, or LED displays, e.g. in calculators, digital instruments, etc. The most common form is a seven-segment display and associated point — *see* Fig. 72. Such a display can light up numerals from 0 to 9, depending on the individual segments energized, with or without the decimal point lighted. Each segment (or point) is, of course, an individual LED.

Specific advantages of LEDs are that they require only low voltages, are fast switching and can be produced in very small sizes, if required. The most widely used seven-segment displays, for example, give figures which are 0.3 in. or 0.5 in. high. Power consumption is relatively low, but an 8-digit seven-segment display could have a maximum power consumption in excess of 2 watts (e.g. $8 \times 7 \times 20$ mA at 2 volts).

This can place restrictions on their application to displays powered by miniature batteries, e.g. digital watches. To provide a reasonable battery life the display is normally left in open circuit and only switched on for the short period when it is required to read the display.

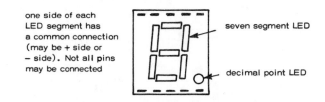

Fig. 72. Typical LED display, as used in calculators, etc. The eight LEDs are internally connected to a common cathode or common anode pin

The *liquid crystal* overcomes this particular limitation since it can be activated by very much lower power (actually a tiny amount of *heat*, which can be produced by an equally tiny amount of electrical energy). Also the display can be made much larger whilst still working at microscopic power levels, so that it can be left 'on' all the time. The liquid crystal has its disadvantages, however. It is far less bright than an LED display, and also suffers from 'dark effect' (like a neon in this respect). Thus to be legible in dim light, the liquid crystal display needs to be illuminated by a separate light course.

Liquid crystal displays operate with low voltage and low current. Current drain can be as little as 1 microamp (1 μA) per segment. A later development, the field effect liquid crystal, can work on even lower voltages drawing microscopic currents (e.g. of the order of 300 nA), again making them an attractive choice for battery powered displays. The field effect liquid crystal also has better contrast, giving a black image on a light background.

CHAPTER 11

Other Components

Separate descriptions of the other types of components likely to be met in electronic circuits are given in this chapter for ease of reference. Many are variations on standard components previously described, but with different working characteristics. The diode family, for example, is particularly numerous.

The Diode Family

The Zener diode is a special type of silicon junction diode which has the particular characteristic that, when reversed bias voltage is applied and increased there comes a point where the diode suddenly acts as a conductor rather than an insulator. The point at which this occurs is called the *breakdown voltage* which, once reached remains constant, even if the negative bias voltage is increased. In other words, once negatively biased to, or beyond, the breakdown voltage, the voltage drop across the diode *remains constant* at its breakdown voltage value, regardless of the actual current flowing through the diode.

This important characteristic makes Zener diodes particularly useful as a source of constant *dc* voltage, or stabilizing a supply voltage, using the type of connection shown in Fig. 73. A series resistor (R) is necessary to limit the amount of current flowing through the diode, otherwise it could be burnt out. Regardless of the value of the input volts, the voltage dropped across the Zener diode

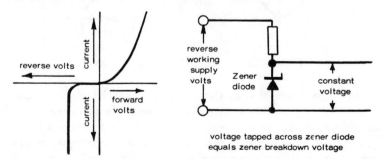

Fig. 73. A Zener diode, working with reverse bias, breaks down at a specific reverse voltage. Connected as shown, it can be used as a source of constant voltage supply. Note the symbol for a Zener diode

will remain constant, so any variations in the input voltage will not affect the output voltage tapped from across the Zener diode. This voltage will be the breakdown voltage of the Zener, which may range from about 2.7 volts up to 100 volts or more, depending on the construction of the Zener diode. If the input voltage falls below the breakdown voltage, of course, the Zener diode will stop conducting and 'break' the circuit.

Performance of a Zener diode as a voltage stabilizing device is limited only by the power rating, which may be quite low, e.g. under 500 mW for the small Zener diodes, but up to 5 watts or more in larger sizes. Its stability will also be affected by the heating effect of the actual current flowing through it, causing a shift in the breakdown voltage so the nominally constant voltage can vary with working temperature. If this is likely to be troublesome (i.e. the type of Zener diode used has a fairly high temperature coefficient of resistance), then connecting two similar diodes in series can greatly improve the temperature coefficient. Also the power rating will be increased.

Another special type of diode is the *Varicap* or *varactor*. These behave as *capacitors* with a high Q (*see* Chapter 6) when biased in the reverse direction, the actual capacitance value being dependent on the bias voltage applied. Typical applications are for the automatic control of tuned circuits or 'electronic tuning', adjusting capacity in the circuit, and thus resonant frequency, in response to changes in signal voltage; automatic frequency control of local oscillator circuits in superhets and TV circuits; and also as frequency doublers and multipliers. Symbols for a Varicap are shown in Fig. 74.

Fig. 74. Alternative symbols for a Varicap

The *tunnel diode* is another type with special characteristics, unlike that of any other semiconductor device. It is constructed like an ordinary diode but the crystal is more heavily doped, resulting in an extremely thin barrier (potential layer). As a consequence, electrons can *tunnel* through this barrier.

This makes the tunnel diode a good conductor with both forward *and* reverse voltage. Behaviour, however, is quite extraordinary when the forward voltage is increased — *see* Fig. 75. Forward current at first rises with increasing forward voltage until it reaches a peak value. With increasing forward voltage it then drops, to reach a minimum or *valley* value. After that it rises again with further increase in forward voltage. Worked in the region from peak voltage to valley voltage, the tunnel diode exhibits *negative resistance* characteristics. Another interesting feature is that any forward current value between peak and valley value is obtainable three times (at three different forward voltages).

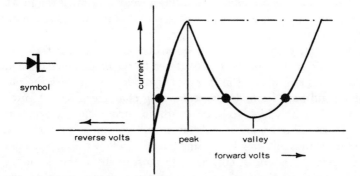

Fig. 75. Symbols (left) and characteristic performance of a tunnel diode. From 'peak' to 'valley' it exhibits 'negative resistance'.

Tunnel diodes have a particular application for very high speed switching, with a particular application to pulse and digital circuitry, e.g. digital computers.

The *Schottky diode* is a metal semiconductor diode, formed by integrated circuit techniques and generally incorporated in ICs as a 'clamp' between base and emitter of a transistor to prevent saturation. Voltage drop across such a diode is less than that of a conventional semiconductor diode for the same forward current. Otherwise its characteristics are similar to that of a germanium diode. A typical circuit employing a Schottky diode is shown in Fig. 76.

For such circuits (i.e. using a Schottky diode as a clamping device associated with a transistor) diode and transistor may be produced at the same time in processing the transistor. This combination device is called a *Schottky transistor (see also* Fig. 76).

Photodiodes
It is a general characteristic of semiconductor diodes that if they are reversed biased and the junction is illuminated, the reverse current

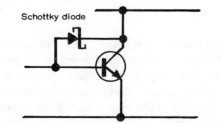

symbol for
Schottky transistor

Fig. 76. Typical circuit for a Schottky diode (left) and the equivalent single component, a Schottky transistor (right)

flow will vary in proportion to the amount of light. This effect is utilized in the *photodiode* which has a clear window through which light can fall on one side of the crystal and across the junction of the P- and N-zones.

In effect, such a diode will work in a circuit as a *variable resistance*, the amount of resistance offered by the diode being dependent on the amount of light falling on the diode. In the dark the photodiode will have normal reverse working characteristics, i.e. provide almost infinitely high resistance with no current flow. At increasing levels of illumination, resistance will become proportionately reduced, thus allowing increasing current to flow through the diode. The actual amount of current is proportionate to the illumination only, provided there is sufficient reverse voltage. In other words, once past the 'knee' of the curve (Fig. 77), the diode current at any level of illumination will not increase substantially with increasing reverse voltage.

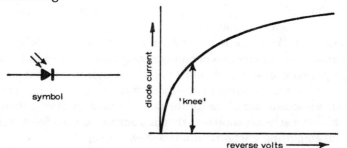

Fig. 77. Symbol (left); and characteristic performance of a photodiode

Photodiodes are extremely useful for working as light operated switches — a simple circuit being shown in Fig. 78. They have a fairly high switching speed, so they can also be used as counters, e.g. counting each interruption of a beam of light as a pulse of current.

There are two other types of light-sensitive diodes — the photovoltaic diode and the light emitting diode (or LED). The *photo-*

voltaic diode generates an *emf* or voltage when illuminated by light, the resulting current produced in an associated circuit being proportional to the intensity of the light. This property is utilized in the construction of light meters. The amount of current produced by a photodiode can be very small, and so some amplification of the current may be introduced in such a circuit. Special types of photodiodes, constructed more like diode valves, are known as *photocells* and are generally more suitable for use as practical light meters.

The *light emitting diode* works the opposite way round to a photodiode-emitting light when a current is passed through it. Light emitting diodes are described in Chapter 10.

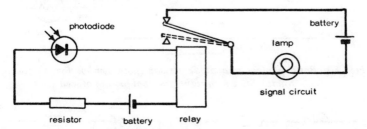

Fig. 78. Photodiode used as a 'light' switch. The rise in current when the diode is illuminated makes the relay pull in, completing an external circuit through the relay contacts

The Phototransistor

The phototransistor is much more sensitive than the photodiode to changes in level of illumination, thus making a better 'switching' device where fairly small changes of level of illumination are present and must be detected. It works both as a photoconductive device and an *amplifier* of the current generated by incident light. A simple circuit employing a phototransistor is shown in Fig. 79.

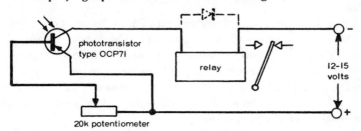

Fig. 79. Practical 'light switch' circuit using a phototransistor. The relay should be of sensitive type and adjusted to pull in at about 2 milliamps. The potentiometer is a sensitivity control. A diode connected across the relay will improve the working of this circuit

A phototransistor and a light emitting diode (*see* Chapter 10) may be combined in a single envelope, such a device being known as an *opto-isolator*. In this case the LED provides the source of illumination to which the phototransistor reacts. It can be used in two working modes — either as a *photodiode* with the emitter of the transistor part left disconnected; or as a *phototransistor* — *see* Fig. 81. In both cases working is governed by the *current* flowing through the LED section.

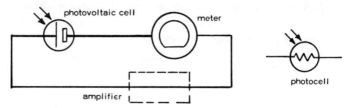

Fig. 80. *Basic photovoltaic diode circuit (note symbol for photovoltaic diode). Shown on right is symbol for a photocell*

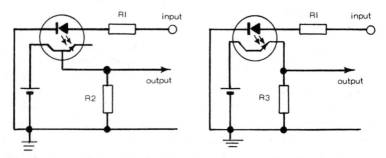

Fig. 81. *Opto-isolator (combined LED and phototransistor), working as a photodiode (left); and phototransistor (right).*

Solar Cells

The photodiode is a *photovoltaic* cell. That is to say, light ('photo') falling on its junction produces a voltage. This voltage measured on open circuit (e.g. with a very high resistance voltmeter connected across the cell), is known as the *photovoltaic potential* of the cell. In this respect it is like a dry battery. Connected to an external load, the cell voltage will fall to some lower value dependent on the resistance in the circuit (*see* Chapter 18).

Photovoltaic cells will develop a potential when illuminated by any source of light. The photovoltaic potential will depend on the construction of the cell, but for any given cell, will be proportional to the intensity of the light.

The solar cell is a photovoltaic cell (silicon photodiode) designed to respond to *sunlight*. Typically it is a small wafer about $\frac{1}{4}$ in. (6.5mm) square, coated with varnish to protect the junction. In 'average' sunlight (about 3000 lumens per sq. ft) its photovoltaic potential is of the order of 500 mV. Connected to a 100-ohm load, it will generate an output current of about 3mA.

To get higher voltages and currents from a *solar battery* a number of cells have to be used connected in series-parallel. Series connection gives a cell voltage equal to the sum of the individual cell voltages. Parallel connection gives a current equal to the sum of the individual cell currents.

Suppose, for example, the solar battery was intended to operate a circuit requiring a nominal 2 volts and give a current of 15 milliamps through a 100 ohm load. From the *voltage* consideration, number of cells required = 2 divided by 500mV (i.e. voltage per cell) = 4 cells.

From *current* considerations, number of cells required = 15 divided by ? (current per cell) = 15 cells.

The solar battery required would thus have to consist of five rows each of four cells, each row consisting of four cells connected in series, and each row being connected together in parallel — Fig. 82.

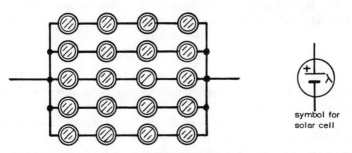

symbol for
solar cell

Fig. 82. Connections for a 2 volt solar battery to give a current of 15 milliamps through a 100 ohm load

A single solar cell can be used to 'measure' solar power, or rather the strength of sunlight at any time. The cell is simply mounted on a suitable panel, sensitive side (negative side) facing outwards and the two cell leads connected to a 0–10 milliammeter. Fig. 83. Directed towards the sun the meter will then give a reading representative of the strength of the sunlight. To measure maximum or peak radiation, point the cell directly towards the sun. To use the instrument as a device for plotting solar energy, called a *radiometer*, the panel should be pointed due south and tilted upwards at an angle approximately 10 degrees more than the local latitude. Readings are then taken at intervals throughout the day, indicating how much solar energy the panel is receiving.

If the meter readings are very low, add a shunt resistor across the meter (shown in broken lines) in Fig. 83 (second diagram). This needs to be a very low value (1 or 2 ohms only). Find a suitable value by trial and error to give near maximum meter reading in the brightest summer sunlight.

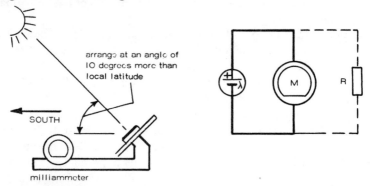

Fig. 83. Aligning a photocell to measure solar energy (left). The circuit on the right is a complete solar energy meter, the meter used being a 0-50 milliamp type

Rectifiers

The conventional diode is a rectifier, its maximum forward current capabilities being determined mainly by its junction area. For signal rectification point contact diodes are usually preferred (*see* Chapter 9), which may limit maximum forward current to 30 to 50 mA, depending on type. Where higher powers are required, larger *rectifier diodes* can be used, with maximum current ratings up to several amps.

In the case of power supplies (*see* Chapter 20), four diodes in 'bridge' configuration are normally used for full wave rectification. Physically this does not mean that four separate diodes have to be connected up — Fig. 49. *Bridge rectifiers* are available as integral units. The average voltage output from such a bridge is 0.9 times the root mean square voltage developed across the secondary of the transformer, less the voltage drop across the rectifier itself.

Metal plate rectifiers are still used for high voltage applications. Each plate is essentially a diode, with a number of plates stacked in series or series-parallel to accommodate the voltage and power loading required. Exactly the same principle can be extended to semiconductor diodes. Selenium rectifiers, which were originally widely used for voltages up to about 100 *rms* have now been virtually replaced by silicon diodes.

Silicon controlled rectifiers or SCRs (also known as *thyristors*) are silicon diodes with an additional electrode called a *gate*. If a bias voltage is applied to the gate to keep it at or near the same potential as the cathode of the diode, the thyristor behaves as if working with reverse voltage with *both* directions of applied voltage, so only a small leakage current will flow. If the gate is biased to be more positive than the cathode, the thyristor behaves as a normal diode. In other words the gate can be used to turn the rectifier on (by positive bias on the gate), thus enabling forward current to be controlled (e.g. preventing forward current flowing over and required portion of a half cycle).

A *triac* is a further variation on this principle, providing bi-directional control. It is virtually a double-ended thyristor which can be triggered with either positive or negative gate pulses.

Structurally an SCR is a four-layer diode, with connections to the inner layers. The terminal connected to the P-region nearest the cathode is the *cathode gate*; and the terminal connected to the N-region nearest the anode the *anode gate* — *see* Fig. 84. Both gates are brought out in a *triac*. Only the cathode gate is brought out in a *thyristor*. Both devices are essentially *ac switches*, the thyristor being effective only on one half of an *ac* voltage, and the triac being effective on both halves.

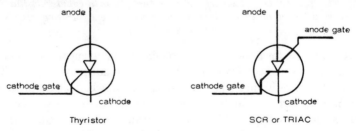

Fig. 84. Symbols for thyristor and SCR

Thermistors

A thermistor is a device designed specifically to exploit the characteristic of many semiconductor materials to show marked reduction in resistance with increasing temperature. This is the opposite effect exhibited by most metal conductors where resistance increases with increasing temperature.

The obvious value of a thermistor is to 'balance' the effect of changes in temperature on component characteristics in a particular circuit, i.e. work as a compensating device by automatic adjustment of its resistance, down (or up), as working temperatures rise (or fall) and resistances of other components rise (or fall). Compensation for temperature changes of as much as 100°C are possible with thermistors — a typical application being shown in Fig. 85. Here the thermistor is used to stabilize the working values of the resistors in an audio amplifier circuit.

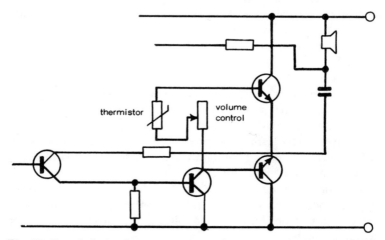

Fig. 85. Practical circuit incorporating a thermistor to counteract fluctuations in value of other resistors in the circuit due to heating effects or temperature changes

Another use for a thermistor is to eliminate current surges when a circuit is switched on 'from cold'. Certain circuits may tend to offer relatively low resistance when first switched on, which could produce a damaging surge of high current. A thermistor in the supply line with a relatively high 'cold' resistance limits the initial peak current surge, its resistance value then dropping appreciably as it warms up so that the voltage dropped across the thermistor under normal working conditions is negligible.

Thermistors are made in the form of rods, looking rather like a carbon rod, sintered from mixtures of metallic oxides. They are not made from the usual semiconductor materials (germanium and silicon) since the characteristics of a thermistor made from these materials would be too sensitive to impurities.

Sensitors are a similar device working in the reverse mode, i.e. their resistance increases with temperature, and vice versa. They thus behave like metallic conductors, although they are made from heavily doped semiconductor material.

CHAPTER 12

Thermionic Valves

Valves are distinctly old-fashioned in these days of transistors and other semiconductor devices, yet they are still widely used in commercial circuits, especially where high power levels are involved.

The basic form of a valve is an evacuated glass envelope containing two electrodes — a cathode and anode. The cathode is heated, causing electrodes to be emitted which are attracted by the anode thus causing current to flow through the valve in the basic circuit shown in Fig. 73 (first diagram). The name *thermionic* valve derives from the fact that heating is the source of electron generation. The Americans call valves 'tubes' or 'vacuum tubes'.

The original form of heating was by a separate low voltage (low tension or LT) supply to a wire *filament* forming the cathode (filament valve, or filament cathode). The later form is a cathode in the form of a tube with a separate heater element passing through it. This is known as an *indirectly* heated cathode (valve), particular advantages being that there is no voltage drop across the cathode (and thus electron emission is more uniform), and also the heating filament can be connected to a separate *ac* supply, if necessary, rather than requiring a separate *dc* supply. Otherwise the working of the valve is identical. Both require a filament (LT) supply and a separate high tension (HT) supply.

The simplest form of valve shown in Fig. 86 is called a *diode*, because it has two internal elements. Its working characteristics are that when the cathode is heated application of HT across the anode

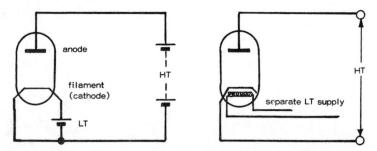

Fig. 86. Diode vacuum tubes (valves) with directly heated filament (left); and indirectly heated filament (right). The filament is the cathode

and cathode will cause a current to flow through the valve, the
current value increasing with anode voltage up to the *saturation
point* — Fig. 87.

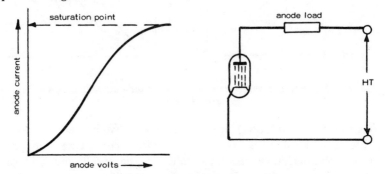

*Fig. 87. Typical diode valve characteristics (left). To do useful work, the
anode current must flow through a load resistor*

Current can only flow in one direction (i.e. HT positive connected
to anode, when current will flow from anode to cathode in the
opposite direction to electron flow (*see* Chapter 2). If the direction of
HT is reversed, then the valve will not conduct. In other words, a
diode works as a *rectifier* in this case.

Note that in this 'working' circuit a *load resistor* is included in the
circuit. Without any external load in circuit all the power input to a
valve would be used up in heating the anode. To do useful work, a
valve must work with a load of some kind or another, so that power is
developed in the load. To work efficiently most of the input power
must do useful work in the load, rather than in heating the anode.
Thus the voltage drop across the load should be much higher than
the voltage drop across the valve.

If a third element, known as a *grid*, is inserted between the
cathode and anode, a negative *bias voltage* can be applied to this to
control the working of the valve and thus the anode current. Such a
valve is known as a triode, a basic circuit for which is shown in Fig.
88, together with characteristic performance curves. The advantage
of this mode of working is that a small change in *bias* voltage (or
voltage applied to the grid) is just as effective as a large change in
anode voltage in bringing about a change in anode current.

The triode is a particularly versatile type of valve which can readily
be made to work as an *amplifier*, or an *oscillator* (an oscillator is
really only an amplifier working with excessive feedback producing

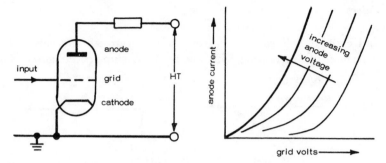

Fig. 88. A triode is a three-element valve. Anode current is controlled by the grid 'bias' voltage

self-sustained oscillation). It does, however, have certain limitations which may be disadvantageous in certain circuits. One is that the inherent capacitance generated between the anode and grid can materially affect the performance of an amplifier circuit where the presence of this capacitance is aggravated by what is called 'Miller effect'. To overcome this particular limitation a positively biased second grid, called a *screen grid* can be inserted between the grid and anode. In effect this acts as an electrostatic shield to prevent capacitive coupling between the grid and anode. Such a four-element valve is known as a *tetrode* — *see* Fig. 89.

Even the tetrode is not without its faults. The 'cure' for one limitation (interelectrode capacitance) has produced another 'fault'. The screen grid tends to attract *secondary emission* electrons bouncing off the anode because it has a positive bias; whereas in the triode the only grid present is negatively biased and tends to repel secondary emission electrons straight back to the anode.

To overcome this effect in a tetrode a fifth element is added, called a *suppressor grid*, inserted between the screen grid and the anode (Fig. 89, second diagram). This acts as a shield to prevent secondary

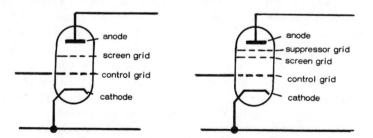

Fig. 89. Tetrode (left) and pentode (right) valves, shown in simple diagrammatic form

emission electrons being attracted by the screen grid. A five-element valve of this type is called a *pentode*.

The Cathode-Ray Tube

The cathode-ray tube has the same number of elements as a triode valve (heated cathode, anode and grid) but works in an entirely different manner. Instead of electrons emitted by the cathode flowing to the anode they are ejected in the form of a narrow stream to impinge on the far end of the tube, which is coated with a luminescent material or phosphor, so producing a point of light. This point of light can be stationary or moving, its direction after emission from the far end of the tube being influenced by the magnetic field created by two additional sets of electrodes or plates, positioned at right angles — Fig. 90. These plates are designated X and Y. Voltage applied to the 'X' plate displaces the light spot in a horizontal direction; and voltage applied to the 'Y' plates displaces the light spot in a vertical direction. (*See also* Chapter 18 for a more detailed description of the working of a cathode ray tube.)

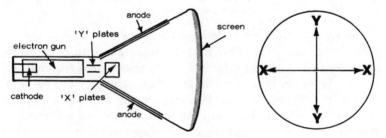

Fig. 90. Simplified diagram of a cathode ray tube

A cathode ray tube can be used as a voltmeter, with the advantage that it puts no load on the circuit being measured. Cathode and anode are connected to a separate supply, the voltage to be measured being connected to the 'Y' plates. A *dc* voltage will displace the spot a proportional distance above (or below) the centreline of the tube. If the voltage applied is *ac*, the light spot will travel up and down at the frequency of the supply which will usually be too fast for the spot to be identified as such, so it will show a trace of light in the form of a vertical line — Fig. 91. The length of this line is proportional to the peak-to-peak voltage of the *ac*.

The more usual application of the cathode ray tube is in an oscilloscope, where a voltage which is increasing at a steady rate is applied to the 'X' plates. If another varying voltage is then applied to the 'Y' plates, the spot will 'draw' a *time graph* of this voltage, or a picture of the *waveform* of that voltage. The 'X' plate varying voltage supply is usually arranged so that once the spot has swept the width of the screen it returns to the start and repeats the picture over and over again.

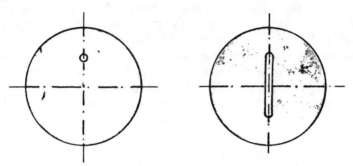

Fig. 91. Simple display of dc *voltage (left) and* ac *voltage (right) on a cathode ray tube. The vertical displacement is a measure of the value of the voltage concerned*

The rate of repetition is determined by the time base of the 'X' plate circuit, this being one of the most important features in oscillo- scope design in order to achieve a steady trace at the required fre- quency. Separate 'shift' controls are also usually provided for both 'X' and 'Y' deflection so that the starting point of the light spot can be set at any point on the screen. Amplifier circuits are also essential in order to be able to adjust the strength input signals applied to the 'Y' plates; and also the 'X' plates if these are also to be fed with an input signal instead of the time base. With these refinements (and others) the cathode-ray oscilloscope is one of the most useful tools an electronics engineer can have.

CHAPTER 13

Integrated Circuits

An integrated circuit or IC consists of a single-crystal 'chip' of silicon on which has been formed resistors, capacitors, diodes and transistors (as required) to make a 'complete' circuit with all necessary interconnections; the whole lot in micro-miniature form e.g. *see* Fig. 92. The cost of an IC chip is surprisingly low, considering how complicated it can be. This is due to the large quantities processed at a time. A 1-inch square 'wafer', for example, may be divided into 400 individual IC chips, *each* containing up to 50 or more separate components. As many as 10 'wafers' may be processed at the same time, yielding 4000 ICs containing the equivalent of nearly a quarter of a million components!

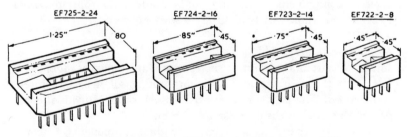

Fig. 92. Some examples of the physical appearance of ICs

Apart from the convenience of having a 'complete' circuit in such a small size, ICs are very reliable because all components are fabricated simultaneously and there are no soldered joints.

Diodes and transistors in an IC chip are formed by exactly the same processes used for producing individual diodes and transistors, but in very much reduced physical size. (Diode and transistor fabrication is explained in some detail in a companion volume by the author, *Building and Designing Transistor Radios, A Beginners*

Guide). Integrated resistors are much simpler. They can be a very tiny area of *sheet* material produced by diffusion in the crystal; or thin film (a millionth of an inch thick) deposited on the silicon diodide layer. Practical resistor ranges which can be achieved are 10 ohms to 50 kilohms, depending on the actual construction, in an area too tiny to see with the naked eye.

Capacitors are a little more difficult. They can either be based on a diode-type formulation (diffused junction capacitor) or again be based on thin-film construction (MOS capacitor). Typical capacitor values achieved are 0.2 pF per *thousandth* of an inch area. Usual maximum values are 400 pF for diffused junction capacitors and 800 pF for MOS capacitors.

Inductances are another story. They cannot — as yet — be produced satisfactorily on silicon substrates using semi-conductor or thin-film techniques. Hence if a circuit specifically needs an inductance in it, the corresponding IC chip is produced without it and an individual inductance is connected externally to the IC.

This, in fact, is common practice in the application of many ICs. The IC is not *absolutely* complete. It contains the bulk of the components, but the final circuit is completed by connecting up additional components externally. It is also usually designed as a multi-purpose circuit with a number of alternative connection points giving access to different parts of the circuit, so that when used with external components connection can be made to appropriate points to produce a whole variety of different working circuits.

Monolithic and Hybrid ICs

Integrated circuits built into a single crystal are known as *monolithic* ICs, with the characteristic of incorporating all necessary interconnections. The problem of electrical isolation of individual components is solved by the processing technique used.

In another type of construction individual components, or complete circuits, are attached to the same substrate but physically separated. Interconnections are then made by bonded wires. This type of construction is known as a *hybrid circuit*.

MSI and LSI

MSI stands for medium-scale integration; and LSI for large-scale integration, referring to the *component density* achieved. The figure of 50 components per chip has already been mentioned as typical of conventional IC construction. These fall into the category of MSI chips, defined as having a component density of more than 12 but not more than 100 components per chip. LSI chips have a much higher component density — as many as 1000 components per chip.

This is largely due to the considerable saving in component sizes possible using thin-film techniques instead of diffusion techniques, particularly in the case of transistors. For example, an MOS transistor can be one tenth the size of a diffused bipolar transistor for the same duty. Hence many more components can be packed into the same size of IC chip.

Op Amps
The operational amplifier or *op amp* is a type of IC used as the basic

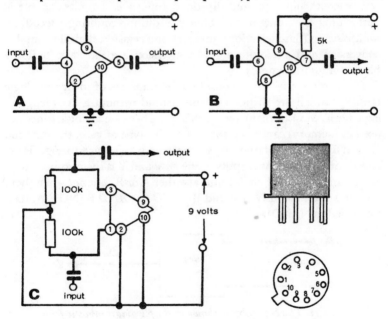

Fig. 93. Three amplifiers based on the CA 3035 IC, all giving a gain of about 100. All external capacitors are 10 µF.

A. Input resistance 2 k ohms

B. Input resistance 670 ohms

C. Input resistance 50 k ohms

D. Enlarged physical drawing of CA 3035. It looks like a typical transistor, but with 10 leads (pins). The LA 3035 'chip' actually incorporates three separate amplifier stages with the equivalent of 10 transistors, 15 resistors and 1 diode. Note how the three different circuits have different connection points into the actual IC circuits building block for numerous *analog* circuits and systems — amplifiers, computers, filters, voltage-to-current or current-to-voltage converters modulators, comparators, waveform generators, etc. It is a typical, almost complete circuit, used in conjunction with a few external components to complete the actual circuit required. Three typical circuits using a simple op amp 'chip' are shown in Fig. 93.

Digital System ICs
Digital systems work in discrete 'steps' or virtually by 'counting' in terms of binary numbers. Basically this calls for the use of logic elements or *gates*, together with a 'memory' unit capable of storing binary numbers, generally called a *flip-flop*. Thus a digital system is constructed from gates and flip-flops. Integrated circuits capable of performing the functions of binary addition, counting, decoding, multiplexing (date selection), memory and register, digital-to-analog conversion and analog-to-digital conversion are the basic building blocks for digital systems.

These give rise to a considerable number of different logic families, which are difficult to understand without a knowledge of logic itself. Most of them are NAND gates because all logic functions (except memory) can be performed by this type of gate, the function of NAND being explained very simply with reference to Fig. 94. A and B are two separate inputs to the *gate* and Y is the output. There will be an output if there is input at either A or B, but *not* when there is input simultaneously at A and B — *NOT* A *AND* B (NOT-AND is simplified to NAND).

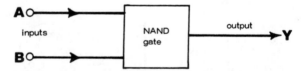

Fig. 94. The NAND gate shown in simple diagrammatic form

The same principle applies with more than two inputs. Further the NAND gate is easily modified to form any of the other logic functions by *negation* or *inversion* modifying the response. These functions are (still restricting description to two inputs):

AND — output when A *and* B input signals are both present

OR — output when input A *or* input B is present

(This is different to a NAND gate for with no input at A or B there is no output; but with a NAND gate there is output)

NOR — NOT-OR

Pursuing the subject of logic could fill the rest of the chapter, or even the whole book, so we will get back to digital integrated circuits.

Digital logic ICs are produced in various different 'families', identified by letters. These letters are an abbreviation of the configuration of the gate circuit employed. The main families are:

TTL (transistor-transistor logic) — the most popular family with a capability for performing a large number of functions. A TTL is a NAND gate.

DTL (diode-transistor logic) — another major family, and again a NAND gate. Its number of functions are more limited.

RTL (resistor-transistor-logic) — a NOR gate which occupies minimum space.

DCTL (direct-coupled-transistor logic) — a NOR gate similar to an RTL, but without the base resistors.

ECL (emitter-coupled logic) — and OR or NAND gate

MOS (metal-oxide-semiconductor logic). Also called a MOSFET since it uses field effect transistors. These chips are of LSI construction, with a very high component density. Some 5000 MOS devices can be accommodated in a chip about 0.15 in. cube. A MOS is a NAND gate.

Regardless of the family used, the basic AND, OR, NAND and NOR gates are combined in one integrated chip of the same family in various combinations of gates and flip-flops to perform specific circuit functions. These functions may or may not be compatible with other families (e.g. TTL functions are compatible with DTL). Also there may be direct equivalents of the complete chip in different families (e.g. TTL, DTL and CMOS). Family development continues and more and more functions are continually appearing, performed by yet more and more ICs appearing on the market.

Analog and Digital Computers

There are two broad classes of computers — *analog* (also spelt analogue) and *digital*. Analog computers work in terms of *voltages* proportional to numerical data. Digital computers work in steps or *binary logic*.

A very elementary form of analog computer is shown in Fig. 95, comprising two input potentiometers and an adjust potentiometer. The adjust potentiometer is used to set the meter reading to full scale deflection when both the input potentiometers are set to zero resistance. Movement of input potentiometer A will then produce a voltage drop in the circuit, shown by a reduction in the meter reading. This reduction will be proportional to the potentiometer movement. The same will happen if potentiometer B is adjusted. It is thus possible to calibrate the meter to indicate the position of either potentiometer A or potentiometer B setting.

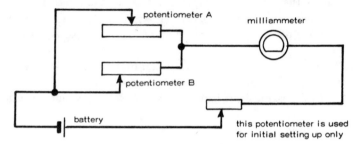

Fig. 95. Elementary analog computer based on two potentiometers

It follows that movement of potentiometer A (or B) can be 'read' on the meter. If both potentiometers are set to mid-position, the adjust potentiometer can be used to set the meter needle in mid position. The scale can then be calibrated to read + values to one side — produced by moving either potentiometer to, say, the right; and — values to the other side — produced by moving either potentiometer to the left. Movement of either potentiometer will thus feed in a + or — signal, indicated by the meter. It follows that the circuit will, in fact, perform *analog addition* and *subtraction* of numerical values fed into it by movement of potentiometers A and B in appropriate directions and by calibrated amounts of movement. The circuit also demonstrates one other major characteristic of an analog computer. The information it can accept, and gives in 'read out' is infinitely variable. Its accuracy does, however, depend on the signal voltages introduced remaining *exactly* proportional to the quantities they represent. In practical analog computer circuits feedback circuits are used to maintain this exact proportionality.

The basic component used in analog computers is an IC (normally represented by a triangular symbol). In the normal mode it is an operational adder (summing amplifier) — Fig. 96, adding the inputs (and taking account of their sign). Factoring can be introduced by adding resistor R1; or direct multiplication or division by connection through a potential divider. Other mathematical processes can also be performed by the op amp — differentiation by replacing R1 with a capacitor; and integration by replacing R2 with a capacitor. The

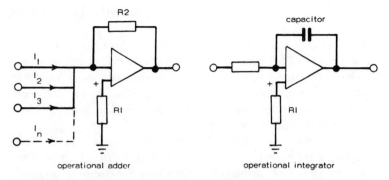

Fig. 96. Operational added (left) and operational integrator (right)

op amp is therefore a most versatile mathematical device, especially associated with other op amps and related circuits. It can also be connected in the inverted mode to change the mathematical sign of an input.

The *digital computer* works in a much simpler way — and at first sight one which is much more limiting (although in fact this is not true). It works basically with bistable circuits which are either 'on' or 'off', and so all mathematical functions are performed in terms of binary numbers. Binary numbers are quite cumbersome when written out, e.g.

0 is   0 in binary notation
1 is   1
2 is   10
3 is   11
4 is 100
5 is 101

already the binary number is getting quite lengthy; and by the time it has reached 100:

100 is 1100100 in binary notation.

Not the sort of numbers to use for mental arithmetic, and when it comes to multiplication and division, multiplication is performed by repeated addition and division by repeated subtraction.

But the system has one enormous advantage, as far as computer working is concerned. The 'reading' of information is dependent only on whether a signal is present ('on') or not ('off'). It does not depend on the *strength* of the signal at all, provided it is not weaker than a certain minimum necessary to trip the bistable circuit. Thus it should be — and is — a very much more precise system than analog working.

The fact that it is apparently cumbersome does not matter at all for computers. The speed at which a bistable circuit can work is so fast that the actual number of individual binary digits or 'bits' which the computer has to process is largely immaterial. And all the operations can be performed by logic *gates* connected in logic circuits to cover exactly the functions required.

Another great advantage of digital workings is that the basic information is simple on-off information which can readily be stored in a memory circuit, to be called back into the circuit at any point required. Storage can be done in *magnetic cores* (literally nothing more than tiny magnetic coils mounted at the crossing points on a grid of horizontal and vertical wires). A 'bit' can be stored at any point in the memory, or recalled from the memory, by signalling that particular grid reference.

Magnetic core memories have the advantage of being cheap, compact and reliable. Once the stored information has been recalled, however, it is lost from the memory (although some magnetic cores do have a facility to retain memory). For permanent storage, or where memory may be required in sequence, magnetic tape is normally used.

CHAPTER 14

Circuit Diagrams

A circuit diagram is a *plan* of a particular circuit showing all the components and all the circuit connections. The components are represented by symbols (*see* Chapter 1), arranged to show all connections simply and clearly, avoiding crossing lines as far as possible. It is a *theoretical* diagram since it does not show the actual size or shape of components, nor their actual position in a built-up circuit. So it has to be redrawn as a *practical* diagram or working plan from which the circuit is actually constructed.

Certain conventions apply in drawing a (theoretical) circuit diagram, but these are not always followed rigidly. The first is that the diagram should 'read' from left to right. That means whatever is 'input' to the circuit should start at the extreme left and be fed through the circuit from left to right. Thus in the simple radio circuit shown in block form in Fig. 97 the input is supplied by the aerial current feeding the tuned circuit, then passing to the detector, then to the amplifier and finally the loudspeaker output. The power supply for the circuit (say a battery) is shown on the far right of the circuit. At first this may seem a contradiction of the rule, if one thinks of the power supply being put into the circuit. But it is not a true *input*; merely a supply to *work* the circuit, and otherwise nothing to do with the circuit. So it is depicted out of the way on the right. There is another good reason for this. Although the supply will feed all the stages 'backwards' in terms of left-to-right reading, it probably will not be required to power the first stage. Hence it is

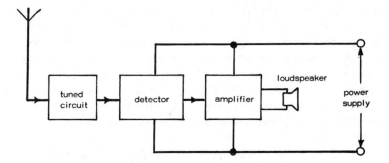

Fig. 97. Basic stages in a simple radio receiver

logical to show feed 'from the right', stopping at the appropriate stage.

Nearly all circuits are based on a *common line* connection, i.e. components in various stages are connected to one side of the supply. This common line is drawn at the bottom of the diagram, as shown in Fig. 98. It is generally referred to as the common earth or earth line, although it may not have any actual connection to an 'earthing' point.

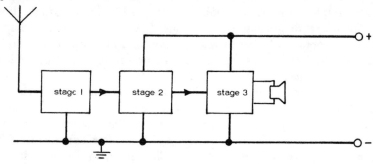

Fig. 98. Stages shown with 'common earth' and power supply connections

A similar common line can also be drawn at the top of the diagram, representing the other side of the supply. Conventionally this top line is the + and the bottom or common earth line the −, although this is not always convenient in designing transistor circuits, so this 'conventional' polarity may be reversed on some diagrams.

Working on this basis and replacing the 'boxes' with individual components, the circuit diagram then looks something like Fig. 99. Each component is identified by a number, or may have its actual value given alongside. Reading' the circuit is fairly straightforward — with a little practice. Starting from the left the 'input' from the aerial is fed to L and C1 forming the tuned circuit. From there it is passed through the diode detector to the amplifier (TR1). TR1 then feeds the final output stage (transistor TR2) driving the loudspeaker. A supply voltage is required only by TR1 and TR2, so the upper 'common' line stops short at TR1 stage. The resistors in the top half regulate the supply; and those in the bottom half (connecting to the common bottom line) establish the working point of the transistors. Additional components (C2 and C3) are required for coupling between stages.

Note that all connecting lines meet at right angles, and where such connection occurs this is further made clear by a • . If a line on the

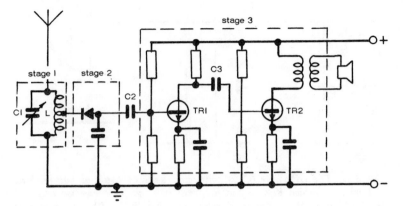

Fig. 99. *The same stages drawn with all components and all connections required*

diagram has to cross another line without any connection to this line, it is simply drawn as a crossing line, as shown in Fig. 101 (*right*). Crossing lines with a ● at the point of meeting indicates that all four lines are, in fact, connected at that point. To avoid possible confusion (or accidentally missing out the ●) connected lines from each side of another line can be drawn as shown in the right hand diagram of Fig. 100.

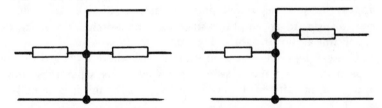

Fig. 100. *Two methods of indicating connection to a common point. That on the right is often preferred for clarity*

So far there should be no confusion at all in 'reading' theoretical diagrams, but they can become more difficult to follow when the circuit becomes more complex or contains a large number of individual components. One common trick-of-the-trade used to avoid too many crossing lines (which could lead to mistakes in following a particular connection) is to 'arrow' a connecting point, or separately designate a common line connection — Fig. 101. Arrowing is usually applied to 'outputs', indicating that this line is connected as an input to a separate stage (or even a separate circuit). Showing a separate 'earth' connection is common with components connected between the 'top' line and 'bottom' line. It indicates clearly that the component is to be

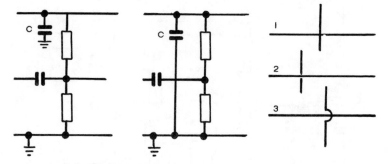

Fig. 101. The capacitor (C) is connected between top and bottom lines, but may be shown in a circuit diagram in either of these two ways. On the right are three ways of drawing crossing lines with no connection. Method 1 is the usual way. Method 2 is clearer; and method 3 clearer still

connected to the 'earth' line, and avoids having to draw this line in close proximity to other components or crossing other lines.

What is less easy to 'read' in terms of actual connections is a circuit incorporating a single physical component which may perform two (or more) separate functions. As a very simple example, a volume control potentiometer for a radio may also incorporate on-off switching. The two *functional* features of this single component may appear in quite separated parts of the circuit, e.g. the volume control prior to an amplifier stage and the switch function in the supply line at the far right — Fig. 102.

This can be even more confusing at first where a ganged tuning capacitor, or a ganged switch is involved, its separate sections appearing in different parts of the circuit although it is actually a

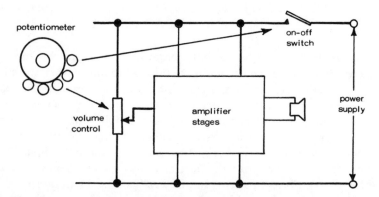

Fig. 102. In this example a single component (the potentiometer connections) appears in two separate parts of the circuit diagram

single physical component. This is the logical — and by far the simplest — way of showing the *theoretical* connections of the circuit; but when it comes to actual construction of the circuit, connections from two or more different parts of the circuit have to be taken to *one* particular component position.

As far as possible actual physical layout of components should follow the same 'flow path' positions as in the theoretical circuit diagram, adjusted as necessary to get components into suitable positions for making connections. Exceptions must arise, particularly as noted above. Just as a theoretical diagram is designed to present the circuit in as simple a manner as possible with all theoretical connections clear, the working circuit must also be planned to be as neat and simple as possible and also as logical as possible as far as placement of components is concerned. It should be prepared as a complete wiring diagram, when it becomes a *working plan*. Almost inevitably it will look more 'confused' than the theoretical diagram, with probably a fair number of crossing wires (unless planned as a printed circuit — *see* Chapter 16) and leads running in various directions. Common connecting points are still indicated by a • , but crossing leads are better shown as definite cross-overs, as in Fig. 103. Then there is less risk of wondering whether or not a • has been missed at that point in preparing the working diagram, particularly as it will be less easy to check connections on a working diagram than on a theoretical circuit diagram.

The theoretical circuit diagram, however, remains the check reference for the working plan — and for checking the circuit when

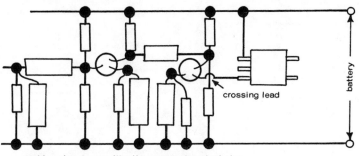

crossing lead

battery

working plan drawn with all components actual size

Fig. 103. Typical component layout drawing or 'working plan'. This is the same circuit as Fig. 99, following the tuned circuit (which would be a ferrite rod aerial)

built. It may also be the only guide available for establishing the correct way round to connect a diode or a polarized capacitor (electrolytic capacitor). Following the direction of current flow (and thus the polarity at any particular point), should be fairly straight-forward, remembering that with a + top line, the direction of current flow will be downwards (from top line to bottom line), through various components on such paths. If the top line is — , the flow direction is obviously reversed. It is also easy to check the direction of current flow through transistors by the arrow on the emitter in the transistor symbol. Direction of current flow *out* of the transistor via the collector will follow in the same direction. Direction of current flow *into* the base of the transistor will be opposite to that of the emitter arrow direction. Fig. 104 should make this clear.

These rules for 'reading' the current flow through transistors should also make it fairly simple to determine the current flow with horizontally connected components on the circuit diagram, and thus establish the correct polarity for electrolytic capacitors appearing in these lines.

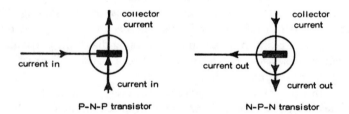

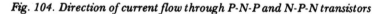

Fig. 104. Direction of current flow through P-N-P and N-P-N transistors

CHAPTER 15

Circuit Construction

One of the big problems facing most beginners is actually how to *construct* a working circuit, i.e. turn a theoretical circuit diagram into a connected-up assembly of components (with all the connections correct, of course!).

The following diagrams show six very simple and straight-forward methods of tackling elementary circuit construction — all capable of giving good results with the minimum of trouble, and especially recommended for absolute beginners at practical electronics.

'Pinboard' Construction (Fig. 105)
Draw out the component layout for the circuit on a piece of thin ply (or even hard balsa sheet), using a ballpoint pen (*not* a lead pencil). Draw in all connections and mark points where common connections occur with a blob (just as on circuit diagrams).

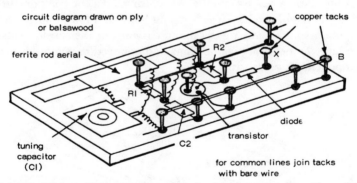

Fig. 105. 'Pinboard' construction starts by drawing a working plan of the circuit on a panel of ply or balsa. Then drive in copper tacks at each connection point. Solder component leads to tacks, and complete circuit as necessary with any additional wiring

Cut out the panel to a suitable size. Drive copper tacks into each 'blob' and simply solder the components in position. Complete additional connections with plain wire.

'Skeleton' Assembly (Fig. 106)
Start again with a component drawing, this time on paper. Lay components in place, bending the leads of resistors and capacitors to

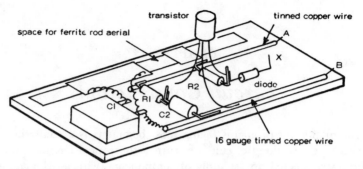

Fig. 106. The same circuit as Fig. 105 tackled by just connecting components together via their open leads, and using bare wire for 'top' and 'bottom' lines

complete connections. Other connections can be complete with two lengths of 16 gauge copper wire for 'top' and 'bottom' common lines. Solder up all the connecting points, adding transistors last. Properly done, such a 'skeleton' assembly can be quite rigid.

Bonded Mounting (Fig. 107)

This is very similar to 'skeleton' assembly except that individual resistors and capacitors are stuck down to a rigid base panel cut from Paxolin sheet. Use 'five-minute' epoxy for gluing the components in place. This will produce a very strong bond in a few minutes. With the main components rigidly mounted you can then bend leads to produce the necessary connections and connection points for other components (e.g. transistors).

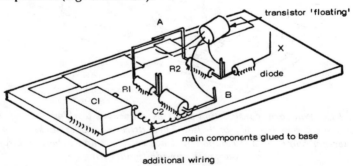

Fig. 107. Bonded mounting is similar to Fig. 106 but all components (except transistors) are glued down to a base panel. Turn up ends of leads to form connecting points

Bus-bar Assembly (Fig. 108)

This is a neat and more 'professional' way to tackle circuit construction. The 'top' and 'bottom' common lines of the circuit are laid

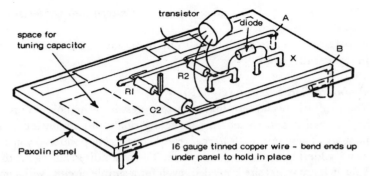

Fig. 108. Lay down the 'top' and 'bottom' lines in tinned copper wire, permanently mounted on the panel. Shorter lengths of bare wire can be used for other common connecting points. Complete by soldering component leads in place

down first by mounting two lengths of 16 gauge tinned copper wire in a sheet of Paxolin, as shown. This will enable most of the resistors and capacitors to be mounted with one lead soldered in place. Complete the rest of the connections as for 'skeleton' assembly.

Tagboard Assembly (Fig. 109)

More durable, and neater, than the previous methods, this involves mounting strips of solder tags (called tag strips) at each of the main connecting points of the circuit layout. These tags can be riveted or bolted through the Paxolin panel. Individual components are then mounted between tags. Any additional connections are formed by short lengths of wire between tags. More time is needed to construct a proper tagboard than with the previous methods, but complete tag-

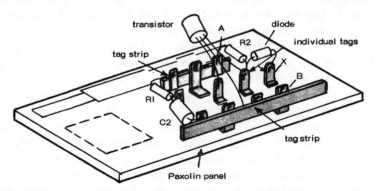

Fig. 109. All the connecting points are formed by tag strips or individual tags mounted on the panel. Solder components to appropriate tags, and complete with additional wiring as necessary

board stops are also available with up to 36 individual tags mounted in two parallel rows.

Pegboard Construction (Fig. 110)

You can buy special terminal pillars to press into the holes in ordinary pegboard and so set up connecting points for mounting components. These terminals have screwed connecting points, so you can avoid soldering components in place. The main disadvantage is that a fairly large panel size is needed, even for a simple circuit, with components well spread out. But it is an easy method of building experimental circuits.

There are various proprietary systems based on variations of the 'pegboard' method. *Veroboards* are Paxolin panels with rows of copper strips, each strip being drilled with a number of holes either 1mm (0.04 in.) or 1.3mm (0.052 in.) diameter. Matching pins (Veropins) can be inserted in appropriate holes to form terminal pillars, and the copper strips cut, as necessary to separate connecting points. Special tools are used for inserting the Veropins and for cutting the copper strips. *Verostrip* is a similar type of board except that the board is narrower (1½ in. wide) and the copper strips run across the board with a break down the centre. Components can be mounted across or along the strips.

Numerous solderless breadboards have also been developed where component leads are simply pushed into the boards where they are held by spring contacts. Contact points are arranged in parallel rows, with either a prearranged pattern of interconnection, or with basic busbar connections on top and bottom rows and others in common groups. Interconnection between groups can be made by wires pushed into spare points in each separate group.

The advantage of such a system is that, apart from avoiding soldering, circuits can easily be modified simply by pulling out a component and plugging into a different position.

For 'permanent' circuits, the beginner will probably find pegboard assembly the best proposition after he has gained some experience in circuit construction — and confidence in being able to draw out component layouts accurately. The ultimate for all forms of compact circuit construction is, however, the printed circuit. Here components are mounted directly through holes in a Paxolin (or glass fibre) panel on which the circuit 'wiring' has been reproduced by etching. This is a separate technique on its own, but easy enough to learn — *see* Chapter 16.

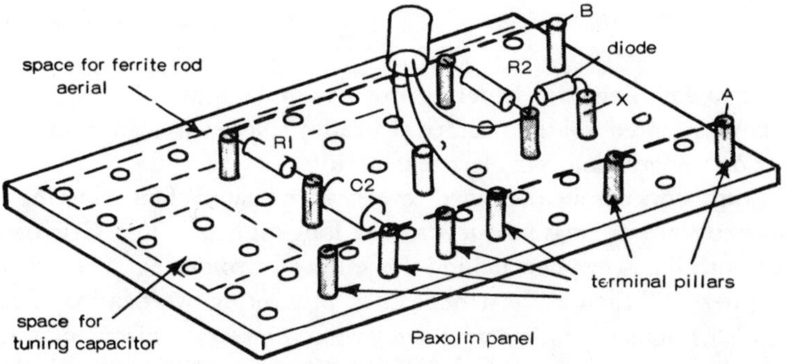
B
diode
space for ferrite rod
aerial
R2
X
A
RI
C2
terminal pillars
space for
tuning capacitor
Paxolin panel

Fig. 110. Pegboard assembly is similar to Fig. 109 except that the board is predrilled to accept terminal posts. Locate these posts at suitable points to act as common connecting points. Connect components to posts, then complete with additional wiring as necessary

Note: all these constructional drawings show the same circuit — a simple single-transistor radio receiver with preamplifier. Components required are:

Ferrite rod aerial with coupling coil
CI — 0–500 pF tuning capacitor
C2— .0I µF
RI — 1 megohm
R2 — 2.2 k ohm
Diode — any germanium crystal diode
Transistor — any rf transistor

The circuit works off a 9 volt battery connected to A and B (polarity depends on whether the transistor used is a P-N-P or N-P-N type). For listening, connect high impedence phones to points X and B.

In case the aerial coil connections are not clear:

(i) The ends of the main coil connect to the two tags on the tuning capacitor (CI).

(ii) One end of the coupling coil connects to the common connecting point of RI and C2

(iii) The other end of the coupling coil connects to the base lead (b) of the transistor

GENERAL RULES

Connections should *always* be soldered, for best results. This applies even on pegboards fitted with screw-type terminal pillars. The one exception to the rule is solderless breadboards where connections are made by spring clips. In all cases, never rely on joints which are formed simply by twisting wires together.

Use a small electric soldering iron for making all soldered joints, and resin-cored solder (electrical grade). Never use an acid type flux on soldered joints in electrical circuits.

Transistors can be damaged by excessive heat. When soldering in place to a circuit, leave the leads quite long (at least 1 in.). Grip each lead with flat nose pliers behind the joint when soldering. The jaws of the pliers will then act as a 'heat sink', preventing overheating of the transistor. Once proficient at soldering, however, this precaution should not be necessary, especially in the case of silicon transistors.

The most common reason why a particular circuit will not work is because *one or more connections have been wrongly made*. This is far more likely to be the cause of the trouble than a faulty component. Always check over all connections after you have made them, using the theoretical circuit diagram as the basic reference. Also with transistor circuits, *be sure* to connect the battery up the right way round (as shown on the circuit diagram).

CHAPTER 16

Printed Circuits

The stock material for making printed circuit boards (PCBs) is copper-clad phenolic resin laminate (SRBP) or glass-fibre laminate. For general use these boards are single-sided (clad on one side only) and nominally 1.5mm thick (about $\frac{1}{16}$ in.).

The procedure for making a PCB involves:

1. cutting the board to the required size and cleaning the copper surface.

2. making a drawing of the conductors required for the circuit on the copper in a *resist* ink

3. etching away uncovered copper areas in a chemical bath

4. removing the resist ink to expose the copper conducting areas or 'lands'.

5. drilling the copper lands ready to take the component leads.

6. degreasing and cleaning the boards as necessary to ensure that the lands 'take' solder readily.

Planning the circuit drawing
Familiarity with the physical size of components to be accommodated on the board is essential, so that holes for leads, etc, can be correctly positioned. There are two ways in which resistors and capacitors can be mounted — horizontally or vertically (Fig. 111). Horizontal mounting is usual for resistors as this reduces lead length to a minimum. Holes are then spaced a sufficient distance apart to allow for easy 90-degree finger-bends on the leads. The same consideration applied to tubular capacitors, mounted horizontally.

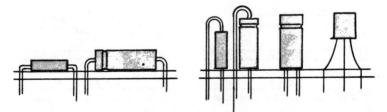

Fig. 111. Horizontal mounting of components on a PC board takes up more space, but is usually more convenient than vertical mounting (except for transistors)

The physical size of capacitors, however, may be much larger than resistors, when vertical mounting may be preferred to save space. Mounting holes then only need to be a little more than half the diameter of the capacitor, matching the position of the top lead taken down the side of the capacitor. Some capacitors have both leads emerging from the same end, especially for vertical mounting on a PCB. Spacing between holes, however, should not be less than twice the thickness of the board (i.e. $\frac{1}{8}$ in.).

Transistors need reasonably wide spacing for their leads. Exceptions are transistor types with leads intended to plug directly into a PCB and certain power transistors needing special mounts. In these cases hole positioning follows the transistor lead geometry. Integrated circuits normally plug directly into matching IC holders, the latter being mounted on the board in holes drilled to match the pin positions.

Layout starts with a tentative design of component positions, sketching in the connections required (i.e. the areas of copper which will eventually form the conducting 'lands'). No connections on a PCB can cross, and a certain amount of trial-and-error sketching is usually needed to achieve this requirement, altering component positions as necessary. If it seems impossible to achieve a complete circuit without crossing connections, then such points can be terminated on the PC drawing on each side of the crossing point, and subsequently completed during assembly of the circuit by bridging with a short length of insulated wire, just as a component normally acts as a bridge between adjacent conductors— Fig. 112.

Having arrived at a suitable layout, with connecting points for component leads indicated by blobs (•), a tracing can then be made of this PCB plan. Certain general rules apply in preparing the final drawing.

1. Conductors should not be less than $\frac{1}{16}$ in. wide.

Fig. 112. If it is (or seems) impossible to avoid a crossing connection on a PC board, 'stop' the crossing lines short (left) and complete the connection with a bridge of insulated wire soldered in place

2. Conductors should be spaced at least $\frac{1}{32}$ in. apart.

3. There should be at least $\frac{1}{32}$ in. between a conductor and the edge of a panel.

4. Bends or junctions in conductors should be radiused or filleted, not sharp-edged.

5. Allow sufficient area of copper around a connecting point so that the copper width at this point is at least twice that of the hole size subsequently to be drilled through it, and preferably more. (Typical 'hole' sizes for miniature resistors, capacitors and transistors are $\frac{1}{32}$ in.).

These points are illustrated in Fig. 113.

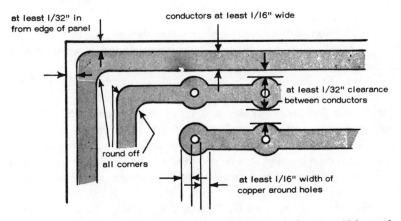

Fig. 113. Basic recommendations for planning conductor widths and spacings on PC boards

It is not necessary to draw all conductors neatly and uniform in thickness. Relatively large 'solid' areas can be left to accommodate a number of common connecting points, simplifying the amount of drawing necessary, e.g. *see* Fig. 114. Large solid areas should, however, be avoided in any part of a circuit carrying high current as this could cause excessive heating of the copper, possibly making it delaminate as it expands. Thus on a PCB for a mains-operated circuit, for example, the maximum area of any particular copper land should not be more than about 1 square inch.

The final drawing is transferred in reverse on to the copper (hence the use of tracing paper). This is because the circuit, as originally planned, shows the *component* side of the board, which is the plain side. Thus the true pattern for the copper side is reversed, mirror-

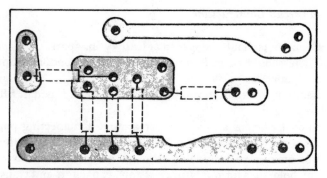

Fig. 114. Quite large areas of copper can be left on low voltage battery powered circuits. There is no need to plan for consistent widths (or shapes) of conductors

image fashion. But before transferring the drawing to the board, the copper surface must be cleaned. This is easily done by washing with detergent and then drying.

A test for cleanliness is to hold the board copper-side up under a tap and let water run on to it. If the water flows freely over the whole area, it is free from grease. If dry patches appear on the copper, these areas are still greasy and require further cleaning.

After tracing the (reversed image) pattern on to the copper, this pattern is then painted in with resist ink or a *resist marker pen* (much easier to use than a brush and ink). Make sure that all the land areas are properly filled, but avoid applying too much ink as this could overrun the required outlines. Finally allow to dry, which should take about 10 to 15 minutes.

The board is then transferred to an *etching bath*. This can consist of a solution of ferris chloride or proprietary printed circuit etching fluid poured into a shallow plastic tray (e.g. the top of a sandwich box). The board is placed in the bath copper side up and left until all traces of copper have disappeared from the surface. Time taken for this will vary with the temperature of the solution and its strength. The process can be speeded up by 'rocking' the bath gently or stirring with a soft brush.

After etching is completed the board is removed, washed under running water to remove any traces of chemical, and dried with a soft cloth. The etching solution can be kept for re-use, if required.

To remove the resist ink, a further liquid known as etch-resist remover should be used. This can be brushed on to the board and then rubbed with a soft cloth, or applied to the cloth first and then rubbed over the board. It should only take a minute or so to remove all the ink, leaving the copper patterns fully uncovered and clean. Wash and dry the board again at this stage to remove any residual traces of etch-resist remover.

Drilling comes next. The following rules are very important:

1. Always drill with the copper side uppermost, i.e. drill through the copper into the board.

2. Always use a sharp drill (preferably a new one).

3. Always use a backing of hard material under the board to prevent the point of the drill tearing a lump out of the back of the board when the point breaks through.

4. 'Spot' the point to be drilled with a small centre punch to prevent the drill running off its correct position when starting to drill.

Use of an electric drill in a vertical drill stand is best for drilling PCBs, and should also ensure 'square' holes. However, because of the small size of drill used, breakage rate of drills can be high if the work is 'pushed' too hard.

The original tracing will now come in handy again for marking the *component* positions on the *plain* side of the board, as a guide for component assembly — Fig. 115. Components are always assembled on the *plain* side, with their leads pushed through their mounting holes until the component is lying flush with the board, e.g. *see* Fig. 111. The exception is transistors, which should be mounted with their leads left fairly long (and preferably each lead insulated with a length of sleeving to prevent accidental shorting if the transistor is displaced).

Before mounting components in position for soldering the copper side should be cleaned again. It will probably have picked up grease marks through handling. An ordinary domestic powder cleaner is best for this, used wet or dry, and rubbed on with a soft cloth. The 'running water' test can again be used as a check for cleanliness, but if the board is wetted, dry with a cloth.

Components are normally soldered in place, one at a time, with their full lead length protruding. Excess length of wire is then cut off as close as possible to the solder. Some people, however, find it easier to cut the leads to length first, then solder.

Provided soldered joints are completed rapidly, e.g. in not more than about 3 seconds, heat-damage to either the board or a component is unlikely. If the iron has to be held in contact with the lead for longer than this, then something is wrong with the soldering technique and heat damage could result, either to the component or by the 'lifting' of the copper land on the PCB. The most likely causes of overheating are using an iron which is not hot enough or too small for the job; attempting to rework a soldered joint which has not 'taken' properly; and trying to remove a lead which has been soldered into the wrong hole.

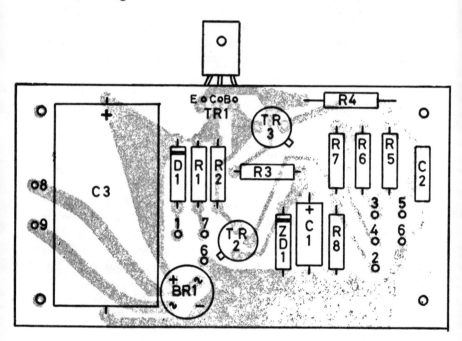

Fig. 115. Example of a printed circuit design with copper area shown shaded and position of components indicated

Simplified Printed Circuit Construction

As a supplement to drawing — and for making neater straight lines — there are available rub-down *transfer sheets* of lines, bends, blobs for connecting points, etc., which can be used to build up the required pattern on the copper, supplemented with ink drawing where necessary. These transfer symbols are resistant to etching fluid, so serve the same function as drawn or painted lines, etc.

It is also possible to buy self-adhesive copper foil precut in the form of lines, bends, etc, similar to transfer strips, but which can be pressed down on to a plain Paxolin panel to complete a printed circuit directly, without the need for any etching treatment. Further shapes can be cut from self-adhesive copper foil blanks. With PC boards made up in this fashion continuous (conductor) sections can be made up from overlapping pieces, provided positive connection is made by solder applied over the joint line.

CHAPTER 17

Radio

Radio broadcasts consist of a *radio frequency* (*rf*) signal generated at a specific frequency allocated to a particular station, on which is superimposed an *audio frequency* (*af*) signal.

Only *rf* will work for *transmission*. The *af* part, which is the actual *sound content* of the signal is, almost literally, carried on the back of the *rf* signal, the two together forming what is called a *modulated* signal.

This combined signal can be produced in two different ways — amplitude or 'up-and-down' modulation, known as *AM*; and frequency modulation (actually a very small variation in *rf* signal frequency about its 'station' frequency), known as FM.

AM is the simpler technique and is widely used for long wave, medium wave and short wave broadcasts. Broadcasting has always been referred to in terms of 'wavelengths' instead of signal frequency, until comparatively recently. The relationship between wavelength and frequency is:

$$\text{wavelength (metres)} = \frac{300,000}{\text{frequency, Hz}}$$

$$\text{frequency, Hz} = \frac{300,000}{\text{wavelength, metres}}$$

(The figure 300,000 is the speed of light in metres/sec, which is the speed at which radio frequency waves travel.)

In the case of FM, very high transmitting frequencies are used — and it is generally referred to as VHF (very high frequency). Actual *wavelengths* are very short, and so it is much more convenient to speak of frequency, the usual range for FM broadcasts being 90–100 *million* Hertz (90–100 MHz). A simple calculation will show that this means a wavelength of about 3.2 to 2.9 metres or say 3 metres.

Regardless of whether the broadcast is AM or FM, though, any radio frequency signal has the same basic requirement for receiving it. The presence of this signal has to be 'found' and then sorted out from signals from other broadcast stations. The 'finding' device is the aerial, and the 'sorting out' device is the *tuned circuit*, which together form the front end of a radio receiver as shown in Fig. 116

(the extreme left-hand part of a circuit diagram — *see also* Chapter 14).

A *tuned circuit* consists, basically, of a coil and variable capacitor, which can be adjusted to show resonance or maximum response to a particular signal frequency applied to it. A full explanation of this behaviour is given in Chapter 7. All the broadcast signals reaching the tuned circuit are very, very weak. Only that signal to which it is *tuned* is magnified or amplified by resonance, so that it stands out at a very much higher level of signal strength.

An actual aerial wire connected to the tuned circuit may or may not be necessary. In the case of AM reception, the coil winding will also act as an efficient aerial wire, if wound on a ferrite rod. This dispenses with the need for an external aerial. The only disadvantage is that the tuned circuit will be *directional*, minimum signal strength being received when the ferrite rod is pointing towards the transmitter sending the signal, and maximum signal strength when the ferrite rod is at right angles to this direction. This effect is most noticeable on small radio receivers which have only moderate amplification (*see* below). To receive certain stations at good listening level, even with maximum adjustment of volume, it may be necessary to adjust the position of the set. Larger receivers normally have enough amplification to compensate for this, but the effect can still be quite noticeable. Also it is always best to operate a receiver below maximum amplification because this minimizes *distortion* of the signal.

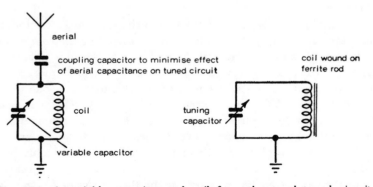

Fig. 116. A variable capacitor and coil form the usual tuned circuit. Strictly speaking this 'tunes' the aerial, if an external aerial wire is used. Most AM receivers use a ferrite rod aerial which does not require an external aerial

The *FM* receiver *does* need an external aerial because a wound coil or a ferrite rod aerial just will not work at this *rf*. For satisfactory results this external aerial also needs to be a special type, known as a dipole, which itself is 'tuned' by making its length one half of the *signal wavelength*. The latter may vary from 11.5 ft. to 9.5 ft. in the 90–100 MHz FM band, so a mean wavelength figure of about 10 feet is adopted, giving a realistic dipole length of 5 feet.

The three practical FM aerial forms are a vertical telescopic aerial extending to 30 in.; a horizontal wire (or rod aerial) with 30 in. 'legs'; or a folded dipole, as shown in Fig. 117.

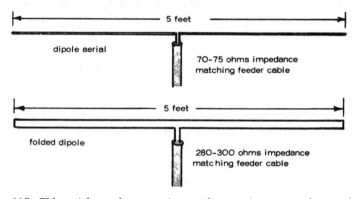

Fig. 117. FM aerials need connecting to the set via a correctly matched feeder cable

Detection

The tuned circuit is much simpler than the foregoing descriptions may appear to imply. It is really a matter of getting the *component values* right, and working with high efficiency (*see also* Chapter 6 and Q-factor). Design of the tuned circuit is a little more complicated when a radio is intended to receive more than one waveband. Even an AM receiver needs separate aerial coils (or at least separated windings on a single ferrite rod) to cover long wave, medium wave and short wave. So the tuned circuit design for an AM receiver could involve three or more tuned circuits selectable by a switch.

In the case of an FM receiver (or the tuning circuit for the FM section of a multi-band receiver), there is really no practical form of wound aerial coil which can be used (a 'theoretical' coil of this type would probably require only a part of a single turn). So the starting point is a dipole aerial. This itself is a 'tuned' circuit (i.e. designed to be resonant with the mean frequency to be covered in the FM band), but its amplification of signal will not be anything like as good as that of the coil-and-capacitor tuned circuit of an AM receiver.

To compensate for this, the FM receiver normally uses an amplifier stage immediately following the aerial, known as a pre-amplifier or *rf amplifier* (because it is an amplifier of signals at radio frequency). This amplified signal is fed to the next stage of the receiver via a tuned output, a typical circuit of this type being shown in Fig. 118.

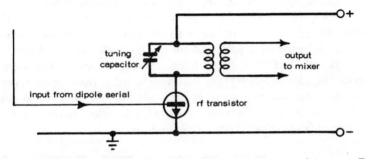

Fig. 118. Basic rf *amplifier (or preamplifier) circuit as used in most FM receivers*

The *detector* stage following the tuned circuit can be extremely simple. In the case of AM working, it only needs to be a diode connected to a potentiometer as its load. This potentiometer also acts as the volume control — Fig. 119.

The signal passed on from the tuned circuit to the detector is a strengthened version of the original *modulated* broadcast signal. In other words, it contains both *af* and *rf*. The *af* part has now done its job in getting the signal into the tuned circuit. Now it needs to be removed, which can be done by rectifying the signal. A diode does this job by 'chopping off' one half of the *rf* signal so that the output from the diode consists of *half-cycles* of *rf*. These half cycles have the *af* content of the signal still imposed, so the next requirement is to filter out the *rf* part to turn the output into an undulating *dc* signal. These undulations follow exactly the same variations as the *af* signal originally imposed on the transmitter *rf* signal at the transmitting station by a microphone, or recording.

As explained in Chapter 6, a resistor and capacitor can act as a filter for any specific frequency required. Thus the diode detector is associated with a matching load (resistance) and associated capacitor forming the required filter circuit, e.g. *see* Fig. 119 — so that only varying *dc* is passed at output from the detector stage. It is usually coupled to the next stage by a capacitor, which has the further effect of 'balancing' the varying *dc* signal about its zero line (i.e. giving it positive and negative values, rather than 'all positive values').

In practice the output load (R in Fig. 119) is usually a variable resistor, which component then also acts as a volume control. The fact that this is followed by a coupling capacitor also avoids any flow of *dc* through the moving contact (wiper) of this control and reduces any tendency to reproduce 'noise' by movement of this control.

The aim in selecting the detector circuit components is that the signal passed by the diode is exactly the same as the original signal generated by the studio microphone (with certain losses and possible distortions!). Fed to a microphone working in reverse (i.e. head-phones or a loudspeaker) they would be heard as the original speech or music, etc. But the signals at this stage are still too weak to have enough power to drive headphones or a loudspeaker, so the next step is to amplify the *rf* signal passed by the detector.

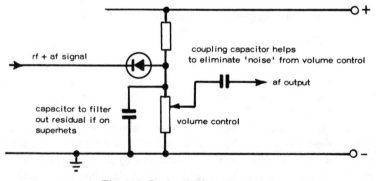

Fig. 119. Basic AM detector circuit

FM Detector

In the case of an FM receiver the detector is a little more compli-cated. It has to detect how the *frequency* of the signal is varying, not its amplitude, so it has to extract the original frequency as well as apply rectification. FM receivers invariably work on the superhet principle, so the frequency to be extracted is the intermediate fre-quency or *if* (*see* below). A basic detector circuit employs a three-winding transformer with primary and secondary tuned to the intermediate frequency (by capacitors C1 and C2 in Fig. 120). The third winding injects a voltage into the secondary circuit, each leg of which carries a diode, D1 and D2, associated with a capacitor C3 and C4.

The working of this circuit is to detect variations in signal frequency in terms of an *af* output, so that the final output is exactly the same, in terms of signal content, as that from an AM detector. Thus it can be dealt with in the same way. The additional components R1 and C5 shown on the diagram are to suppress unwanted signals which may be present after detection.

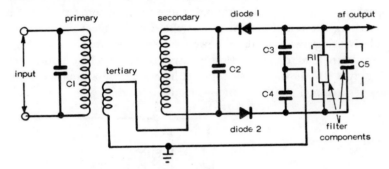

Fig. 120. Basic FM detector circuit

Amplifier Stage(s)

A single transistor can provide amplification of signal strength up to 100 times or more (*see* Chapter 9 for typical amplifier circuits). The main requirement of the amplifier following the detector is that the transistor be specifically suitable for amplifying *af* (when it can act as a further block to any residual *rf* remaining in the input signal to the amplifier). Ideally there should be no *rf* signal present at the input to the amplifier stage (it should have been filtered out in the detector stage), since any *rf* voltage presented to the amplifier stage could cause overloading.

Theoretically, at least, any amount of amplification can be produced by adding additional transistor-amplifier stages (Fig. 121). This does, however, greatly increase the chances of distorting the signal, so there are practical limits to the number of stages which are acceptable in simple circuits. Much better results can be produced by more sophisticated circuits, particularly the superhet (*see* below), where first some intermediate signal is amplified *before* detection; and subsequently amplified again *after* detection.

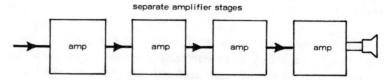

separate amplifier stages

Fig. 121. Adding amplifier stages is not necessarily a good thing for amplifying an af signal as each stage can amplify distortion produced in the previous stage

Output Stage

The *af* amplifier output (or last *af* amplifier stage output if more than one stage is used) can develop enough power to drive a loudspeaker, as in Fig. 121, although there may be some problem in matching the (amplifier) output to the (loudspeaker) input, particularly using low to medium power transistors which require a high impedance load to match. Most loudspeakers have a load impedance of only 4–16 ohms. A basic solution here is to use an output transformer to match the different load characteristics, e.g. as in Fig. 122.

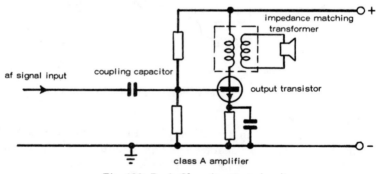

Fig. 122. Basic Class A output circuit

This relatively simple solution does, however, have one particular limitation (for the more technically minded, it is called a Class A output). It is relatively inefficient and so draws a high current in providing a suitable listening level from the loudspeaker. It is satisfactory for use in car radios, but represents too heavy a current drain for most other battery powered receivers.

These normally use a Class B output circuit where the last amplifier transistor 'drives' a pair of transistors which effectively work in a 'push-pull' circuit operating the loudspeaker. The output power obtained is considerably more than double the power available from a single transistor; also an output transformer is not necessary. Most modern *af* amplifiers for radios end up in a push-pull output stage of this type, like the circuit shown in Fig. 123.

The limitations of simple radio receivers are mainly connected with the limitation of a detector. A detector is most effective working with an *rf* input voltage of 1 volt or more. Signals derived directly from an aerial circuit are seldom more than a few millivolts in strength, and the weaker the signal the less effectively they will be detected in any case. In other words, the range of stations that can be picked up is limited, and no amount of amplification *after* detection can make up for this limitation.

This limitation or lack of *sensitivity* can be overcome by amplifying the incoming signal *before* detection, so that the detector is always working with good signal strength. This can be done by *rf amplification* of the aerial signal by introducing an amplifier stage right at the beginning of the circuit as in the FM receiver (Fig. 118); or by the superhet working. The latter also improves the *selectivity* of a receiver, or its ability to tune in sharply to wanted signals and reject nearby station signals.

The Superhet
Having arrived at a 'standardized' output stage, it is equally true to say that nearly all modern radio receivers are of the *superhet* type,

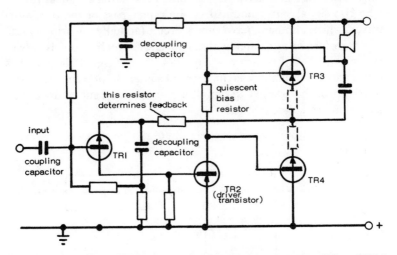

Fig. 123. Basic Class B output circuit. TRI works as a preamplifier. TR2 is the driver. TR3 and TR4 are a complementary pair of transistors, working alternatively in 'push-pull'. The two resistors shown by dashed lines may be added to improve the stability of the circuit. These only need to be of very low value (e.g. 1 ohm)

which is considerably more complicated than the circuit traced through above. The whole 'front end' works on an entirely different principle.

Starting point is the tuned circuit (ferrite rod aerial) in the case of an AM receiver; or a dipole aerial feeding an *rf* amplifier in the case of a FM receiver (the latter amplifying the modulated radio signal in conjunction with a tuned resonant circuit). In both cases the boosted 'tuned' signal is fed to an *oscillator-mixer*.

This is a two function circuit, although its duty is usually performed by a single transistor associated with a tuned *oscillator* circuit. This tuned circuit is mechanically coupled to the aerial tuning in the form of a ganged capacitor (i.e. two separate variable capacitors coupled or 'ganged' to move together when the tuning control is adjusted), so that it 'tracks' the aerial circuit tuning whilst remaining separated from it *by a constant frequency*. This difference is known as the *intermediate frequency* or *if*, and is usually 470 kHz above the aerial frequency (it can have other values in certain sets; and can also be below rather than above the aerial frequency).

The oscillator part of the oscillator-mixer is concerned with generating this fixed intermediate frequency tracking exactly above (or below) the signal frequency to which the aerial circuit is tuned. The two signals are combined in the mixer part of the oscillator-mixer, which also has a *fixed* tuned circuit (actually the primary side of a transformer associated with a capacitor) which responds only to the intermediate frequency — Fig. 124. This *if* signal also now has the same *af* modulation as the original signal. In other words it is a duplicate of the 'wanted' *af* signal, but at this stage superimposed on a fixed intermediate frequency.

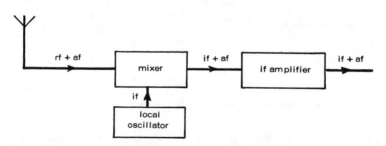

Fig. 124. *'Front end' of a superhet receiver, showing how the incoming* rf *plus* af *signal is transformed into an* af *signal now imposed on a fixed intermediate frequency* (if). *This makes amplification without distortion more simple to achieve*

There are a number of technical advantages to this seemingly unnecessary complication of incoming signal treatment. First, the process of superheterodyning gives much better *selectivity* or rejection of unwanted signals. Then the signal output from the mixer is at a constant frequency, making it easy to amplify with the further possibility of eliminating any remaining unwanted frequencies since an *if* amplifier has fixed — and virtually exact — tuning.

In practice, *if* amplification is usually carried out in two stages (AM receivers) or three stages (FM receivers). The detector then follows *after* the *if* amplifier stages — Fig. 125. Each *if* amplifier stage consists of a tuned transformer, adjustment of tuning being done by an iron dust core in the transformer coil former. Once correctly adjusted, in setting up the receiver for proper working, no further adjustment is necessary, or made. The cores are sealed in this position.

The remainder of the radio circuit follows as before — *rf* amplifier stage(s) following the detector, terminating in (usually) a push-pull output stage. But there is just one further refinement which can be added. By feeding a proportion of the signal passed by the detector back to the first *if* amplifier stage, automatic volume control (normally called automatic gain control or *agc*) can be achieved. If the signal strength passed by the detector starts to rise to a point where it could become distorted, then feedback via the *agc* line automatically reduces the amount of signal entering at this point, so maintaining the detector working under optimum conditions.

Agc applies only to the control of amplification of signal in the *if* amplifier stages. The output or gain of the final *rf* amplifier stage(s)

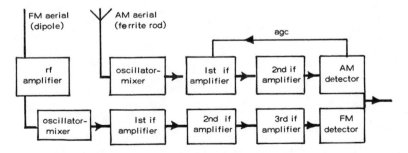

Fig. 125. Basic design of an AM/FM receiver shown in block form. The only common circuit is an af *amplifier (usually a Class B output) following the detector*

is governed by a separate volume control (potentiometer), typically located before the first *af* amplifier stage. This potentiometer, incidentally, usually has a capacitor connected in parallel with it to filter out any residual *if* which may have got past the detector.

CHAPTER 18

Television

Television makes special use of a *cathode ray tube*, which in turn has certain characteristics in common with a thermionic valve (*see* Chapter 12). It has a heater; a cathode which emits electrons; an anode to which electrons are attracted, and a control grid. Unlike a thermionic valve, however, the electrons are directed at the enlarged end of the tube *or screen* which is coated with a *phosphor* material.

It is, in fact, a special type of cathode ray tube (*see* Chapter 12). The narrow end of the tube acts as an electron 'gun', shooting electrons past the anode section. Electrons impinging on the screen generate more electrons which are attracted back to the anode, equivalent in effect to each electron reaching the screen being bounced back to the anode. Thus no electrons, and hence no charge, actually collects on the screen. Meantime, however, each electron reaching the screen makes the phosphor glow, which glow persists for a short period after the electron has been 'bounced back'.

The brightness of the glow produced is dependent on the type of phosphor (which also governs the colour of the glow), and the strength of bombardment by electrons. The latter is controlled by the bias voltage applied to the grid. In other words, grid bias adjustment is the brightness control on a TV tube — Fig. 126. The actual brightness is also enhanced by an extremely thin layer of aluminium deposited over the phosphor to act rather like an outward-facing mirror, but transparent from the other side as far as electrons are concerned.

To produce a *picture* from electron bombardment, two other controls are necessary. The first is a means of deflecting the electron beam so that a single spot can trace out a particular path covering all the variations in picture density over the whole screen area. The second is a means of focusing the electron beam into a tiny spot so that the traced 'picture' is sharp, not fuzzy.

Deflection is achieved by directing the stream of electrons through two sets of parallel coils set at right angles to each other like the X and Y deflection plates in a simple cathode ray tube (Chapter 12). Signal voltages applied to the X-coils will deflect the beam sideways; signal voltages applied to the Y-coils will deflect the beam vertically. Combined X and Y signals will thus direct the beam towards any spot

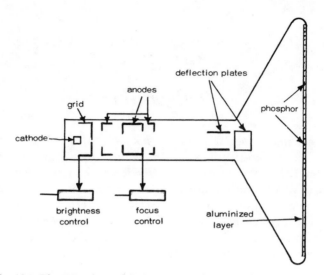

Fig. 126. The TV tube and its basic controls, shown in simplified form

on the screen, depending on the resultant effect of the two signals.

Focusing, meantime, is achieved by using supplementary cylindrical anodes arranged to work as an electronic lens, with the focusing effect adjustable by varying the voltage applied to one (or more) of these anodes. These anodes come before the deflection plates, i.e. in the parallel or 'gun' section of the tube rather than in the divergent section.

Electronic circuits can respond very rapidly — which is how television can be made to work at all! To 'paint' a picture on the screen a spot of light (produced by a focused electron beam) has to traverse the whole picture area, zig-zag fashion, at least 25 times per second if the picture is to appear reasonably free of flicker. It does this in a number of parallel lines, usually running from left to right, with rapid 'flyback' between lines — Fig. 127. The greater the number of lines the clearer the picture will be, i.e. the better the *definition*. The standard commonly adopted is 625 lines (per second). So the actual frequency at which lines appear, called the *line frequency*, is $25 \times 6255 = 15625$ per second. This line pattern is known as a *raster*. The lines making up the raster can actually be seen if you examine a television screen close up; or turn up the brightness control when no picture is being transmitted. Only the parallel lines will be seen in the raster. During flyback the cathode ray tube is cut off and no lines appear on the screen.

In practice the picture is scanned 50 times per second, not 25. This is fast enough to eliminate any trace of flicker, but using an optical 'trick' the actual *picture frequency* is still only 25 per second. Scanning takes place in two stages — first the odd lines only, then the even lines. Each scan therefore builds up only half the picture, the two halves following each other to present the complete picture.

Movement of the lines downwards is accomplished by the *time base* circuit starting with the first (odd) line and restarting a line at the left *two* positions down each time. This continues until the scanning has reached $625 \div 2 = 312\frac{1}{2}$ lines. The spot then flies back to the top again, starting half way along the first even line and repeats the process to scan the $312\frac{1}{2}$ even lines which make up the second half of the picture. This process is known as *interlacing*. Actually a few lines get left out in this changeover process, but this does not show up on the picture.

Picture transmission and picture reception operate in reverse mode. The television camera scans the scene to be transmitted in 625 lines at a picture frequency of 25 per second, and turns the light spot response into electrical signals. The number of lines has been quoted as governing picture definition, but this is not the whole story. 625

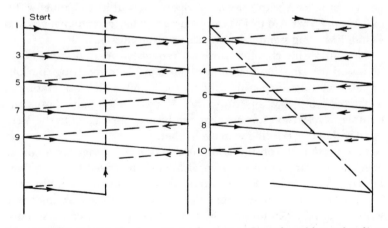

Fig. 127. Illustrating the formation of a raster. First the odd number lines are scanned from left to right (solid lines) with 'flyback' between each line (dashed lines). Downward movement is controlled by the time base. After scanning half the picture lines, the time base flies back to the top. All the even number lines are then scanned, with 'flyback' between each line (right hand diagram). At the end of the last line the time base flies back to top left to start the sequence all over again. These two diagrams superimposed would represent a complete raster.

lines gives good picture definition from top to bottom, e.g. the picture is built up top-to-bottom from 625 'strips'. There is also the question of how many individual picture elements are covered by each strip. The answer is about 600 as an absolute minimum for good picture definition side-to-side or the equivalent of 600 phosphor 'dots' making up each line. The total number of individual 'dots' or picture elements in each whole picture trace is thus 375,000. Since the picture frequency is 25 times per second, this calls for a transmitted signal frequency of 2.5 MHz.

These 'intelligence' signals are broadcast like any other radio transmission — superimposed on a *carrier wave* to produce a modulated signal which can be picked up by a receiver and decoded on a similar principle to ordinary radio reception, except that the decoder now has to handle radio frequencies of 2.5 MHz and not *af* signals, so the TV receiver decoder is considerably more complicated than a radio receiver detector.

There is also another important difference. Carrier wave frequencies have to be much higher than modulation frequencies for satisfactory results. Hence the frequency of television picture signals is of the order of 40–45 MHz (or in the VHF range). It is invariably an AM broadcast with sidebands, operating within a channel width of 8 MHz. Either AM or FM can be used for the accompanying sound signal, FM being standard with 625-line transmissions.

The fairly wide channel width or frequency spread occupied by a television transmission does make it susceptible to receiving spurious signals upsetting the picture (but not the sound, which is operating in a narrow band like any ordinary FM receiver). It also limits the number of television stations that can be accommodated in the VHF band without interfering with each other.

This particular consideration also makes the design of a colour television system even more complicated than it need be using first-principle electronics. For example, it would mean expanding the bandwidth to three times its black-and-white figure to transmit three separate pictures simultaneously in the three primary colours. Since this is not an acceptable solution, colour information has to be contained within the 8 MHz channel allowed for black-and-white transmissions, which becomes an extremely complicated subject and virtually impossible to describe in simple terms. Strangely enough, however, it does simplify one other problem — the essential requirement that a colour television should also be able to receive black-and-white transmissions in black and white. The broadcast stations still have the opposite problem of ensuring that colour transmissions can be received on black-and-white sets in black-and-white!

The conventional colour TV tube is made with three guns, one for each colour — red, green, and blue; with each dot on the screen formed by separate red, green and blue phosphors arranged in a triangle. The picture is thus scanned by a *triangle* of beams converging on each *triangle* of dots at the same rate as in a black-and-white picture. The resulting sharpness of the colour picture depends primarily on the accuracy of convergence, and on some tubes may vary noticeably from centre to edge and/or top to bottom. This may be a limitation of that particular design of tube and associated circuitry, or merely a matter of convergence adjustment (which is usually factory-set and can be quite complex). There are improvements in this respect in the case of some modern tubes, e.g. simplifying convergence problems by using in-line guns and colour spots.

CHAPTER 19

Batteries

Batteries may be divided into two main groups:

1. *Primary batteries*, where the electrochemical action is irreversible (once the battery has been discharged it cannot be re-energized);

2. *Secondary batteries*, where the electrochemical action is reversible and they can be charged and discharged repeatedly (but not indefinitely), to be used over and over again.

Primary batteries are popularly called *dry batteries*; and secondary batteries are called *accumulators*. These descriptions are not strictly correct, but they are convenient and widely accepted. In fact, all primary batteries have some form of 'wet' electrolyte (usually in the form of a paste or jelly); and many secondary batteries may have so-called 'dry' cells (implying the use of a non-liquid electrolyte). And just to show how general classifications cannot always be strictly correct, some types of primary (non-rechargeable) battery systems are, in practice, rechargeable!

Dry Batteries (Primary Batteries)

A battery is made up of one or more individual *cells*. Each cell will develop a specific nominal voltage, depending on the battery 'system'. To build a battery with a higher voltage than that given by a single cell, two or more cells are connected in series.

The nominal cell voltage given by different 'systems' is:

(i) carbon-zinc — 1.5 volts

(ii) alkaline-manganese — 1.5 volts

(iii) mercury — 1.35 or 1.4 volts (depending on the depolarizer used)

(iv) silver oxide — 1.5 volts

(v) zinc-air (a comparatively undeveloped system as yet) — 1.2 volts

Choice of the best 'system' is dependent on cost and utilization of the battery. Carbon-zinc dry batteries are by far the most popular because they are relatively cheap and readily available in a wide range of voltages and sizes, the latter governing the *capacity* of the battery or the amount of electrical energy it can store.

Typical construction of a carbon-zinc cell is shown in Fig. 128. Higher voltage dry batteries may be produced by connecting two or more cells in series in a suitable 'pack'; or more usually by employing layer type construction, as shown in the second diagram.

The two main disadvantages of standard carbon-zinc cells (or batteries) are:

1. Under load (i.e. when current is being drawn from them), they are subject to a build-up of internal resistance due to 'polarization' which has the effect of causing the output voltage to drop. The more rapid the rate of discharge, the more pronounced this voltage-drop effect.

2. The outer zinc case (normally enclosed in a card jacket) is continually being eaten away by chemical attack, which accelerates as the cell becomes more and more discharged. If the battery is left in situ in a torch, radio, battery holder, etc, and left in a near-discharged or discharged state, the zinc casing will soon be eaten right through, allowing corrosive electrolyte to escape.

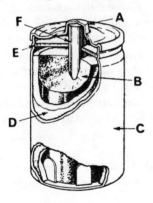

KEY
A Brass Cap
B Manganese Dioxide Depolarizer
C Zinc Case (Negative Plate)
D Ammonium Chloride Electrolyte
E Cardwasher
F Carbon Rod (Positive Plate)

KEY
A Metal Top Cap
B Plastic Top Cover
C Carbon Rod
D Soft Bitumen Sub-seal
E Seal
F Tamping Washer
G Depolarizer
H Paper Lining
J Zinc Cup
K Paper Tube
L Metal Jacket
M Bottom Paper
N Metal Bottom Cover

Fig. 128. Standard carbon-zinc cell (left) and HP cell (right)

To a large extent this can be avoided by using leakproof construction where the zinc case and cell bottom are enclosed by metal covers (usually tinplate). Even the *leakproof* carbon-zinc battery can suffer casing failure and leak if left in place when discharged, however, for there is always a tendency for the contents of the cell to swell and burst the casing when the cell is in a discharged state.

The *high power* or HP cell is an attempt to overcome both limitations. It is invariably of leakproof construction and also employs a more effective depolarizing agent. As a result it has a superior performance under high current drains, and also increased capacity. It is also more expensive, so an HP cell is not always a 'best buy'. In fact, the additional cost of HP cells is really only justified where the current drain is higher than that recommended for standard carbon-zinc cells. And if positive 'leakproof' performance is essential, another 'system' is better (in particular, alkaline-manganese).

Table 1 is a useful guide both to the nominal voltage of standard sizes of carbon zinc batteries and their recommended operating range in terms of *current drain*. Capacities cannot be quoted for carbon-zinc dry batteries since the capacity of a particular size and type of cell can vary widely with different current drains, and the actual operating cycle. This capacity — or 'battery life' — can really only be determined by practical experience in that particular usage.

Alkaline-Manganese Cells
Construction of a typical cylindrical alkaline-manganese cell is shown in Fig. 129. They are generally available only as single (1.5

KEY
A Inner Steel Case and Positive Terminal
B Outer Steel Case
C Insulating Disc
D Insulating Spacer
E Absorbent Sleeve
F Electrolyte in Absorbent Material
G Zinc Anode Cylinders
H Depolarizer Cylinders
J Electrolyte Immobiliser
K Sealing and Insulating Grommet
L Steel Double Plate Negative Terminal
M Battery Jacket

Fig. 129. Typical construction of a modern alkaline-manganese cell

volt) cells, and in a much more restricted range of sizes (*see* Table II). They are considerably more costly than carbon-zinc cells, but have three main advantages:

1. Voltage drop under load is far less pronounced.

2. They have a long, almost indefinite 'shelf life' (i.e. there is virtually no loss of capacity in storage with fresh cells, and even with partly-discharged cells).

3. They have far more capacity for the same size compared with

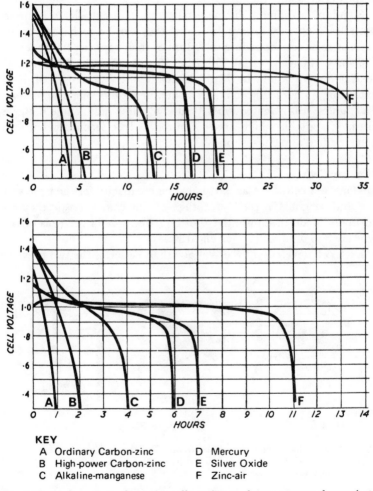

KEY

A Ordinary Carbon-zinc   D Mercury
B High-power Carbon-zinc   E Silver Oxide
C Alkaline-manganese   F Zinc-air

*Fig. 130. Performance of AA size cells under moderate current drawn (top)
and higher current drawn (right)*

carbon-zinc cells, and thus a longer battery life under similar working conditions (size for size, about six times that of a standard carbon-zinc cell).

With this type of battery 'system' it is also possible to quote actual capacities, as given in Table III. They can be regarded as a direct substitute for equivalent carbon-zinc cells where their advantages outweigh the higher cost. They are particularly suitable for constant heavy-load (i.e. high current drain) duties for which the carbon-zinc battery is least suited.

Standard alkaline-manganese batteries — are basically primary batteries. Contrary to popular belief they *can* be recharged at *low* current rates corresponding to 1/10th of their capacity figure, e.g. a 750 mAh capacity cell at 750/10 = 75 mA. Full charge time with such a current would be 12–14 hours. There *is* a danger of a cell exploding if charged at too high a rate, or for too long a time, however, unless it is a vented type. Vented alkaline-manganese batteries are now available which are recognized as true secondary batteries and specified as rechargeable.

Mercury Cells
Mercury dry cells have an extremely high capacity for their physical size and weight. To take advantage of this characteristic they are normally made in diminutive 'button' sizes, but larger cells of this type are also produced — *see* Fig. 131.

Each cell develops a lower voltage than either of the two previous

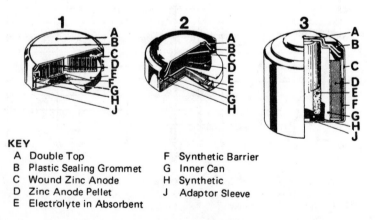

KEY
A Double Top   F Synthetic Barrier
B Plastic Sealing Grommet   G Inner Can
C Wound Zinc Anode   H Synthetic
D Zinc Anode Pellet   J Adaptor Sleeve
E Electrolyte in Absorbent

Fig. 131. Typical construction of mercury cells 1. wound anode 2. pellet-type 3. cylindrical type

types described, but this voltage is largely unaffected by load (current drain). Thus, after a slight initial drop, a mercury cell will continue to maintain a *constant* voltage until almost fully discharged, when the voltage will drop off quite rapidly.

This can be a considerable advantage in many applications. It can also be a disadvantage for there is no means of assessing how far a mercury cell has been discharged by measuring its 'on load' voltage. It simply comes to the end of its useful life suddenly.

It also has some other minor disadvantage (apart from high cost!). There is an appreciable drop in capacity if operated at temperatures below about 5° C (40° F), and at freezing point the cell becomes more or less inert, although it will recover again on warming up. It is also not suitable for use in contact with copper or copper alloy, so contacts used with mercury batteries should be nickel plated (or stainless steel).

Silver-Oxide Cells
Silver-oxide cells have similar characteristics to mercury cells, with a higher nominal cell voltage which remains substantially constant under load. They also have superior low temperature characteristics, but cost is higher than a mercury cell and availability limited.

Accumulators (Secondary Batteries)
The familiar lead-acid accumulator was widely used in the early days of radio as a low tension battery. The only type of rechargeable battery which has a significant application in present-day electronics is the modern nickel-cadmium battery. It is the one type of secondary battery system in which 'gassing' can be eliminated on charging, so it can be constructed in fully sealed cell form (although many types are, in fact, constructed with re-sealing vents as a precaution).

Nickel-cadmium batteries offer numerous advantages — no deterioration during storage in either charged or discharged state (except that a charged cell will suffer a loss of about 20 per cent of its capacity a month); charge/recharge cycle life of at least several hundred and possibly thousands (depending on actual use); suitability for high discharge rates (and high charge rates with vented cells); robustness; wide operating temperature range (−40° C to + 60° C); and suitability for operating in almost any environment.

Disadvantages are high initial cost (which is generally recoverable in long cycling life); and the fact that the nominal voltage per cell is only 1.2 volts. However, the discharge characteristics are substantially 'flat', i.e. output voltage is maintained regardless of load. To ensure a 'full' recharge being given, the favoured technique is first to discharge the cell or battery fully under controlled conditions and then give a full charge.

Construction of a typical nickel-cadmium 'button' cell is shown in Fig. 132. Button cells are produced with capacities ranging from about 0.1 Ah up to 1.75 Ah—*see* Table IV. Battery 'packs' are made

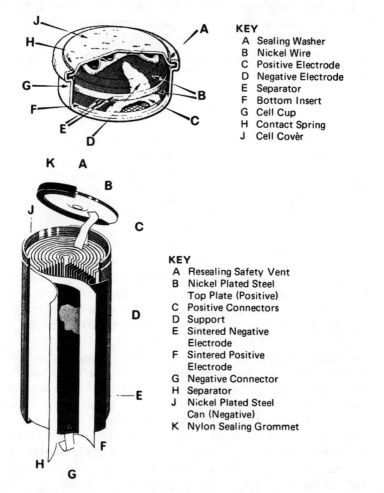

KEY
A Sealing Washer
B Nickel Wire
C Positive Electrode
D Negative Electrode
E Separator
F Bottom Insert
G Cell Cup
H Contact Spring
J Cell Cover

KEY
A Resealing Safety Vent
B Nickel Plated Steel
  Top Plate (Positive)
C Positive Connectors
D Support
E Sintered Negative
  Electrode
F Sintered Positive
  Electrode
G Negative Connector
H Separator
J Nickel Plated Steel
  Can (Negative)
K Nylon Sealing Grommet

Fig. 132. Typical nickel-cadmium button cell (top) and cylindrical cell (right)

up by welding separate cells together in a stack of the required number (the whole stack then usually being given with a shrunk-fitted plastic sleeve) — larger capacities are provided by cylindrical and rectangular nickel-cadmium cells.

Vents are fitted to all nickel-cadmium cells intended for use with high current drains and/or high discharge rates. These vents are of the re-sealing type so that under normal operating conditions the cell is virtually a sealed type.

A problem with nickel-cadmium cells can be that it is virtually impossible to solder connections to the cells. However, cells (and battery packs) are produced with tags spot-welded to each end (and at individual tapping points on a battery, if required) to which soldered connections can be made. Without such tags, connections to a cell or battery pack would normally have to be made via spring contacts.

Specifications for nickel-cadmium batteries may quote capacities at either the 10-hour or 5-hour (discharge) rate. The difference can be significant, as the effective capacity of a given battery is dependent on discharge rate — higher discharge rates resulting in a loss of capacity. Equally, any maximum current rating for a nickel-cadmium battery can be somewhat arbitrary. Most cells can withstand higher current drains than the specified maximum, but continual cycling under such working conditions can substantially reduce the actual number of useful life cycles. Special nickel-cadmium cells with sintered electrodes are designed for operation with very high current drains.

General Battery Rules

1. To increase the *voltage* of a battery, increase the number of cells connected in series to make up the battery. For a battery of given voltage:

$$\text{No. of cells required} = \frac{\text{required battery voltage}}{\text{volts per cell}}$$

If the number of cells so calculated is not a whole number, use the next whole number up. For example:

Battery voltage required is 9 volts

Cells to be used are nickel-cadmium, volts per cell 1.2.

$$\text{No. of cells required} = \frac{9}{1.2}$$

$$= 7.5 \text{ cells}$$

Therefore make up the battery from 8 cells connected in series. (The actual battery voltage will then be $8 \times 1.2 = 9.6$ volts. If necessary the additional voltage can always be 'dropped' in a circuit, e.g. using a dropping resistor).

2. To increase the *capacity* of a battery, connect two (or more) batteries of the required voltage in parallel. Two batteries connected in parallel will *halve* the current drain from each battery, thus doubling the capacity. Basically, in fact, the capacity of the original battery is multiplied by the number of similar batteries connected in parallel.

TABLE 1. STANDARD CARBON-ZINC BATTERIES
(up to and including 9 volts)

Type*	Nominal voltage	Weight (gr)	Size (millimetres) Length or diameter	Width	Height	Recommended current range (mA)
U12 (Pencell)		25	14·3	—	50	20–30
U11		45				20–60
SP11		45	26	—	50	20–60
HP11		45				0–1000
U2	1·5	90				10–50
SP2		90	34	—	60	10–50
HP2		90				0–2000
AD4		600	67	67	102	100–250
FLAG		880	67	—	166	100–250
1839	3		26		100	20–60
PP11		450	65	52·5	91	10–100
126	4·5	370	103	35	91	0–250
AD28		450	100	35	106	30–300
481		1130	113	66	165	100–250
PP1		283	65	56	56	5–50
PP8	6	1100	65	52	200	20–150
996		580	67	67	102	30–300
991		1500	136	72	125	30–500
PP3		38	26·8	17·5	48	0–10
PP4		51	25·4	—	50	0–10
PP6		142	36	35	70	2·6–15
PP7	9	200	46	46	62	5–20
PP9		425	66	52	81	5–50
PP10		1250	65	52	225	15–150
PP11		459	65	52	91	5–50

Type numbers shown in **bold** are World standards. Other types numbers are Ever Ready. See Table II for equivalents.

TABLE 11A. CARBON-ZINC BATTERY EQUIVALENTS

	UM1	UM2	UM3	UM4	UM5
UNITED KINGDOM					
Ever Ready	U2 HP2 SP2	U11 HP11 SP11 C11	HP7 D14	U16 HP16	D23
Exide (Drydex)	T2 SP2 HP2	T15 SP11 HP11	HP7 T5	U16 HP16	DL33
Ray-O-Vac	D2 RR13 2LP HC2	RR14 1LP C1 HC1	7R RR15	–	–
Vidor	HP2 V2 LPV2 VT12	HP11 V11 LPV11 VT13	V12 HP14 VT14	HP16 V16	–
CONTINENTAL					
Berec	U2 LPU2 SP2 PP12	U11 SP11 PP13	U7 U12 PP14 PP15	U16	D23
Cipel (Mazda)	GT1 F20 P20 LF20	MT1 RFM LF14	AC1 LF6	RFB	PA1 PC1
Daimon	253 289 250 251	258 259 287	298 296	291 294	292 295
Hellesens	VII-33 VII-34 VII-36 VII-37	VII-24 VII-25 VII-26 VII-27	VII-28 VII-38 VII-75 18	17	14
Leclanché (France)	T1 R20 R20S	R14 R14S	R6 R6S HA6	RO3	R1
Leclanché (Suisse)	208	207 604	201 202 601 602	–	–
P. L. B.	E2 E4 E7	E5	E15 E18	–	–
Pile Wonder	EXPOR MARIN AMIRO	BABIX JUNON ESCAL	SONAT VEBER NAVAL	EXTAZ	SAFIR
Superpila	60	61	63 433 AC7	68	67
Varta (Pertrix)	211 212 222 232	214 235 236 213 233	244 251 284 280	239	245 249
Witte Kat	667	668	666	–	665
JAPAN	UM1	UM2	UM3	UM4	UM5

TABLE 11B. EQUIVALENT 1·5 VOLT DRY CELLS

Alkaline-Manganese (Mallory)	Carbon-Zinc (Ever Ready)	International
Mn-9100	D23	N
Mn-2400	U16/HP16	AAA
Mn-1500	U12 (pencell)	AA
Mn-1400	U11/SP11/HP11	C
Mn-1300	U2/SP2/HP2	D

TABLE III. ALKALINE-MANGANESE CELLS (1·5 V Nominal Voltage)
(Mallory 'Duracells')

Mallory Type No.	Diameter		Height		Weight			Capacity mAh	Rated at mA
	in.	mm	in.	mm	oz.	gr.			
Mn-625G	0·610	15·5	0·238	6·1	0·106	3		125	2·5
Mn-825	0·905	23	0·228	5·8	0·245	7		300	1
Mn-1	0·625	15·9	0·645	16·4	0·336	9·5		580	15
Mn-9100*	0·455	11·6	1·130	28·7	0·340	9·6		580	15
Mn-2400	0·405	10·3	1·745	44·3	0·400	11·3		750	25
Mn-1500	0·555	14·1	1·960	50	0·820	23·2		1500	50
Mn-1400	1·020	25·9	1·940	49·3	2·340	66·5		5000	83
Mn-1300	1·315	33·4	2·377	60·4	5·030	142		10000	250

* Reversed polarity Midget cell (the top end is negative and the bottom end is positive).

TABLE IV. BUTTON-TYPE NICKEL-CADMIUM CELLS
(1·2 V per cell)

Type	Capacity Ah	Diameter in.	Diameter mm	Thickness in.	Thickness mm	Weight oz.	Weight g
DEAC 225	0·225	1	25	0·36	9·0	0·5	14·25
500	0·50	1·35	34·4	0·39	9·7	1·0	28·5
Ever Ready NCB 9	0·09	0·9	22·7	0·21	5·2	0·23	6·5
NCB 20	0·20	1·0	24·8	0·29	7·4	0·39	11·0
NCB 28	0·28	1·35	34·4	0·21	5·3	0·58	16·5
NCB 55	0·55	1·35	34·4	0·37	9·45	1·00	28·5
NCB 90	0·90	2·0	50·5	0·33	8·3	2·26	64
NCB 175	1·75	2·0	50·7	0·59	14·9	3·53	100

CHAPTER 20

Power Supplies and Charges

The application of a step-down transformer associated with a bridge rectifier and a smoothing capacitor to provide a low voltage *dc* supply from an *ac* mains supply has already been described in Fig. 49, Chapter 8. Rather more sophisticated circuits may be used where it is desirable to ensure a stable *dc* voltage, e.g. for operating an FM transistor radio from the mains instead of a battery.

A circuit of this type is shown in Fig. 133, the component values specified giving a stabilized *dc* output of 6 volts from 240 *ac* mains. The bridge rectifier following the transformer provides full wave rectification, smoothed by capacitor C1 in the conventional manner.

The input to output voltage is dropped across transistor TR1. The emitter voltage of transistor TR2 is set by the Zener diode ZD at 2.7

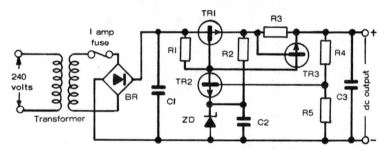

Fig. 133. Stabilized output dc *charger for working off mains components*
R1 — 680 ohms
R2 — 820 ohms
R3 —   1 ohm
R4 — 39 ohms for 6 volts dc output
*   180 ohms for 7.5 volts dc output*
*   330 ohms for 9 volts dc output*
R5 — 560 ohms
C1 — 1000 µF or higher value
C2 — 10 µF
C3 — 0.01 µF
T — 240/15 volt transformer
BR — bridge rectifier
TR1 — BD131
TR2 — BC108C
TR3 — BC108C
ZD — Zener diode BZYbbC2V7

volts. The output voltage is divided by R4, R5 and when the voltage across R5 is about 3.2 volts, TR2 begins to conduct. This diverts some of the current flowing through R1 into the base of TR1 so that TR1 starts to turn off. Thus since the base current, and thus the voltage drop across the collector-emitter junction of TR1 is controlled by TR2, the output voltage is stopped from going any higher than the design voltage.

Conversely, if a heavy load is applied to the output it will tend to cause a drop in output voltage and so also the voltage on the base of TR2 will tend to fall. The effect of this is that TR2 will start to turn off, allowing more current to flow into the base of TR1 which turns on to maintain the output voltage.

Voltage stabilization will be maintained until the output current rises to the order of 4–500 mA. At this point the voltage across R3 becomes greater than the turn-on voltage of TR3, which will start to conduct. This will tap current from the base of TR1, causing TR1 to start to turn off, hereby reducing the output voltage, so that the current does not rise any further. In other words the circuit is automatically protected against overload, even down to short circuit conditions. In the latter case the voltage will fall to almost zero, with the current still being maintained at 400–500 milliamps. The capacitors C2 and C3 are not strictly necessary, but are additional smoothing components.

The circuit can be adapted to provide a number of different *dc* output voltages, selectable by switching. To do this resistor R4 is replaced by a chain of resistors R4A, R4B, R4C, as in Fig. 134. The

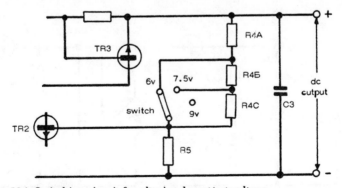

Fig. 134. Switching circuit for altering dc *output voltage*
R4A — 39 ohms
R4B — 150 ohms
R4C — 150 ohms

values given, together with the previous circuit component values, will provide selectable outputs of 6 volts, 7.5 volts, or 9 volts, with voltage stabilized in each case up to a maximum current drain of 400–500 milliamps.

Battery Chargers

Either a bridge rectifier circuit, or the voltage-stabilized circuit just described can also be used for battery charging. In this case smoothing is not so important as the presence of a certain amount of 'ripple' in the *dc* is held to be beneficial for charging. Normally, however, at least one smoothing capacitor would be desirable in the charger circuit.

Since it is not necessarily evident whether a charger is working or not, an indicator lamp or ammeter is normally desirable in a charger circuit. A lamp merely indicates that the charger is 'on' and the output circuit is 'working'. It can be tapped directly across the circuit at any point, the preferred form of lamp being an LED since this draws minimal current, although a small filament bulb will do as well. An LED needs to be associated with a ballast resistor to drop the necessary voltage at this point; a filament bulb does not, but is needed to work as a voltage dropper to a 6 volt bulb to be used in this circuit. (*See* Fig. 135.) Note that an indicator lamp on the mains side of the transformer, on the secondary side between the transformer and the bridge rectifier, would not necessarily confirm that the output was 'working' with an output load connected.

In the case of a meter indicator, this would simply be an ammeter

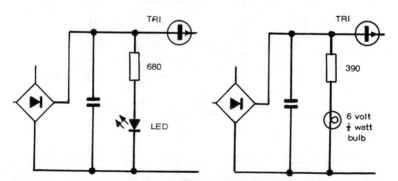

Fig. 135. Alternative arrangements for charging indicator lamps. A bulb of appropriate voltage could be used in one of the output leads without a dropping resistor

(or milliammeter, as appropriate) connected in series in one or other of the output lines.

dc Input Chargers

There is also a call for chargers which can charge low voltage batteries direct from another battery, such as a 12-volt car battery. In this case, since the input is *dc*, a transformer cannot be used to 'set' the required voltage, nor is a rectifier necessary.

Fig. 136 shows a charger circuit designed to provide a stabilized 6-volt *dc* output (charging voltage) from a 12-volt input supply. Essentially it is the same as that of Fig. 133 without the transformer and rectifier, but a diode is included to protect the transistors in the circuit against reverse voltages. Working of the circuit is the same as that described previously, with automatic short-circuit protection. Like the previous mains circuit, too, it can be adapted to provide a range of output voltages, using exactly the same values for the chain of resistors as in Fig. 134.

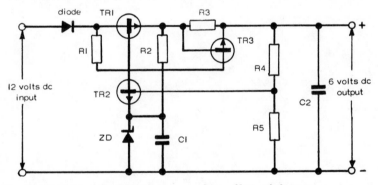

Fig. 136. Stabilized 6 volt dc *charger working off 12 volt battery*
Components R1 — 680 ohms
              R2 — 820 ohms
              R3 — 1 ohm
              R4 — 39 ohms
              R5 — 560 ohms
              C1 — 10 μF
              C2 — 0.01 μF
              TR1 — BC131
              TR2 — BC108C
              TR3 — BC108C
              ZD — Zener diode BZY88C2V7
              diode — IN4001 (or equivalent)

Index

Index